AF183280

WIRELESS COMMUNICATIONS AND NETWORKS - RECENT ADVANCES

WIRELESS COMMUNICATIONS AND NETWORKS - RECENT ADVANCES

Editor

Carlos Granger

scitus
academics

Wireless Communications and Networks - Recent Advances

Edited by **Carlos Granger**

ISBN: 978-1-68117-610-9
Library of Congress Control Number: 2016946465

© 2017 by
SCITUS Academics LLC,
616, Corporate Way, Suite 2, 4766,
Valley Cottage, NY 10989

www.scitusacademics.com

This book contains information obtained from highly regarded resources. Copyright for individual articles remains with the authors as indicated. All chapters are distributed under the terms of the Creative Commons Attribution License, which permits unrestricted use, distribution, and reproduction in any medium, provided the original author and source are credited.

Notice

Reasonable efforts have been made to publish reliable data and views articulated in the chapters are those of the individual contributors, and not necessarily those of the editors or publishers. Editors or publishers are not responsible for the accuracy of the information in the published chapters or consequences of their use. The publisher believes no responsibility for any damage or grievance to the persons or property arising out of the use of any materials, instructions, methods or thoughts in the book. The editors and the publisher have attempted to trace the copyright holders of all material reproduced in this publication and apologize to copyright holders if permission has not been obtained. If any copyright holder has not been acknowledged, please write to us so we may rectify.

Printed in United States of America on Acid Free Paper

PREFACE

Wireless communication is one of the greatest contributions of technology to humanity. Wireless communication means the transmission of information over a distance without the assistance of wires, cables or other types of electrical conductors. The distance transmitted can be anywhere between several meters (e.g. remote control of a TV) and thousands of kilometers (e.g. radio). Wireless technology in networking and telecommunications has become the most exciting field as the rapid growth of mobile phone usage in various satellite services and wireless services today. Internet and wireless LANs bring about enormous networking and telecommunications changes.

This book equips readers with comprehensive information covering basics, open issues and recent developments in wireless communications and networks. This guide focuses on the hottest issues from probably the lowest levels to the top of wireless communication networks and presents "real time" research results on these issues. The book aims to serve as an invaluable tool for research strategists, engineers, researchers, faculty members, scientists, educators and students in these rapidly evolving fields and to encourage them to explore these vast, exciting and rapidly evolving research fields actively.

TABLE OF CONTENTS

1

ANT COLONY-BASED ENERGY CONTROL ROUTING PROTOCOL FOR MOBILE AD HOC NETWORKS UNDER DIFFERENT NODE MOBILITY MODELS

Jipeng Zhou, Haisheng Tan, Yuhui Deng, Lin Cui and Deng Deng Liu

Department of Computer Science, Jinan University, Guangzhou, China

ABSTRACT

The energy of nodes is limited in mobile ad hoc networks (MANETs). In order to extend the network lifetime, how to select the best route is a critical issue for routing protocols in MANETs. In this work, we propose the ant colony-based energy control routing (ACECR) protocol to find an optimal route by using the positive feedback character of ant colony optimization (ACO). In our ACECR protocol, the routing choice depends on not only the number of hops between nodes and the node energy, but also the average and the minimum energy of the routes. The performance of our ACECR routing protocols is evaluated indifferent mobility models. In addition, we do extensive simulations to study the movement characteristics of different mobility models and their effect on routing protocols. Simulation results show that ACECR has a better performance in balanced energy consumption and a longer network lifetime compared with existing protocols.

KEYWORDS

Mobile ad hoc network; Routing protocols; Ant colony optimization; Energy Mobility model

1. INTRODUCTION

A MANET is self-organizing and of a dynamic topology, which enables wireless communication among mobile devices without relying on a fixed infrastructure. Due to limited resources such as power, bandwidth, processing capability, and storage space at the nodes as well as mobility, it is important to reduce routing overheads in MANETs, while ensuring a high rate of packet delivery. Since the battery of nodes is limited, the energy of nodes and the life time of network is a critical problem in MANETs.

Ant Colony optimization (ACO) [1] is a computational model of swarm intelligence which provides efficient solutions to some optimization problems. In ACO routing algorithms, multiple ants created by a node traverse the network to search paths between two nodes. If the ant finds a path, it lays down pheromone on the path. The amount of pheromone depends on the quality of the path such as its number of hops, delay, and energy of nodes on the path. A data packet is transmitted on a link with probability based on the amount of pheromone. ACO routing algorithm exhibits interesting properties for MANETs, as it works in a fully distributed way and provides multi-path routing.

Many routing protocols [2, 3, 4, 5, 6, 7, 8, 9, 10] have been studies in networks, routing algorithms for MANETs based on ACO have been proposed in [2, 3, 4, 5, 6, 7, 8]. AntHocNet [2] is a hybrid multi-path algorithm with the principle of ACO-based routing in MANETs. It uses forward ant (Fant) to find routes and backward ant (BANT) to build routes from the source node to the destination. In recent years, the power problem in MANETs has been receiving significant attention in mobile nodes. Power management schemes have two objectives [4], which are to minimize the total power consumption in the network and to minimize the power consumption per node. The first method targets to extend the overall network lifetime and the latter aims to extend individual node's lifetime. The overall power consumption reduction can be achieved by two different approaches. An ant-based on-demand energy route (AOER) protocol is proposed for mesh networks [5]. Compared to other ant-based route protocols, AOER needs less memory storages and lower processing capabilities, because the structures of ants can be simplified by the specific inverse pheromone table. An energy-aware ant-based routing (EAAR) protocol is proposed in [6]. It takes into account various factors such as the power consumed in transmitting a packet and the residual battery capacity of a node, so that they can increase the battery life of the nodes by reducing the repetitive use of a subset of nodes. The minimum battery energy remaining from the weakest node and the hop-count of the route are used as the metrics for path discovery. An ant-based energy-efficient routing protocol (AEERP) is proposed in MANETs [7], where the route choice is dependent on not only the number of hops between nodes, but also the energy consumed in transmitting packets and the residual energy of nodes. AEERP can balance the energy consumption of nodes in the network and extends the network lifetime.

Generally, a mobility model is designed to describe the movement pattern of nodes in MANETS [11], and to specify their locations, velocities, and accelerations over time. The mobility model is one of the most important factors in evaluating the performance of a routing protocol in a MANET. In this work, we evaluate the effect of existing mobility models on ant colony-based energy control routing protocols through simulations. The rest of this paper is organized as follows. In section 2, we propose ant colony-based energy control routing protocol ACECR. In section 3, we give some simulation results. Section 4 concludes the paper.

2. Ant colony-based energy control routing protocol

The efficient foraging behavior of naturally occurring small-sized and energy-constrained ants is studied in the theory of ACO [1]. ACO uses the concept of artificial ants, which is analogous to the natural ants that behave as packets in MANETs. In ACO-based routing algorithms, pheromone content is used to choose the best paths out of a given network. It can be used to forward data stochastically. Data for the same destination can be spread over multiple paths with more data transmitted on higher quality paths, which results in load balancing. ACO-based routing algorithms perform better in many ways due to their proactive and iterative behavior. These kinds of algorithms also reduce variability and errors in networks by choosing a trusted path which have behaved well for quite some time.

2.1 Data structures of ants

In this section, we propose an ant colony-based energy control routing protocol ACECR. In ACECR, when a source node wants to send a data packet to its destination, it checks its pheromone table and finds the next relay node in the path. If the pheromone table does not have next node to the destination node, the source node will start a path discovery process. The source node sends out a request packet, which is called Fant (forward ant). When a node receives Fant, it will update the node list and record the node which Fant has passed. Each node in networks forwards the Fant packet until it reaches the destination. When Fant arrives the destination node, it will create a new packet which is called Bant (backward ant). The destination will send the Bant back to the source node along the reverse route. Structures of Fant and Bant are shown in Tables 1 and 2, where SID stands for the source ID, DID is the destination ID, Seq is the sequence number of the forward ant (backward ant), HOP is the number of hops from the source node to the current node for Fant or the number of hops from the destination node to the current node for Bant, the path is the listed of node IDs of routing path, E_{min} is the minimum residual energy of nodes in the path, E_{sum} is the summation of residual energy of nodes in the path, and TTL is the living time of the ant.

Table 1. Structure of Fant

SID	DID	Seq	HOP	Path	TTL

Table 2. Structure of Bant

SID	DID	Seq	HOP	Path	E_{min}	E_{sum}

To maintain the amount of pheromone on a link, each node has a pheromone table that stores the amount of pheromone on each incident link. The pheromone table, as shown in Table 3, is a two-dimensional array, in which the row and the column denote neighboring nodes and destination nodes, respectively. A value ϕ_{nd} in the pheromone table of node *u* is the amount of pheromone on a link (*u,n*) in paths to destination *d*. Thus, the amount ϕ_{nd} of pheromone represents how good the link (*u,n*) is to transmit a data packet to the destination *d*. The notations used in this paper are shown in Table 4.

Table 3. Structure of pheromone table at node *u*

	Destination nodes					
Neighbor	d_1	d_2	...	d_i	...	d_n
n_1	$\phi_{n_1 d_1}$	$\phi_{n_1 d_2}$	...	$\phi_{n_1 d_i}$	...	$\phi_{n_1 d_n}$
n_2	$\phi_{n_2 d_1}$	$\phi_{n_2 d_2}$	...	$\phi_{n_2 d_i}$	...	$\phi_{n_2 d_n}$
...	...	...	...	...	...	...
n_k	$\phi_{n_k d_1}$	$\phi_{n_k d_2}$	...	$\phi_{n_k d_i}$	...	$\phi_{n_k d_n}$

2.2 Route discovery process

To establish a path from a source *s* to a destination *d*, source s creates a Fant and broadcasts it to all neighbors of *s*. The aim of Fant is to search a path from source *s* to destination *d*, by traversing the network and establishes the pheromone track to the source node. Node *i* forwards a Fant according to Procedure send Fant (*i*), if node *i* has routing information available for *d*, the node *i* will forward it to next node *j* with probability $P_i(j) = \frac{(\phi_{jd})^\beta}{\sum_{s \in N_i} (\phi_{sd})^\beta}$, where N_i is the neighbor node set of node *i*, and $\beta$is a constant parameter. If *i* has no pheromone for the destination *d* (i.e., $\forall j \in N_i, \phi_{jd}=0$), Fant is broadcasted to all neighbors of node *i*.

```
 /* Node i sends a Fant*/
Procedure sendFant(i)
1. if i is not a destination, then
2.   if there is an entry for
destination d in pheromone table, then
3.     it selects a next node j with
probability Pi(j)
       and it sends Fant to node j
      else
4.     Broadcast Fant;
      endif
   else
5.    sendBant(i)
   endif
endProcedure
```

Table 4. Notations used in this paper

Symbol	Comments
Fant	Forward ant
Bant	Backward ant
ϕ	Pheromone of nodes
E_{avg}	The average energy of a path
E_{sum}	The total energy of a path
E_{min}	The minimum energy of a path
$E_{max}(i)$	The maximum energy of node i
$E_{cur}(i)$	The residual energy of node i
$N(i)$	The neighbor node set of node i
HOP	Hop count of a path
$S_{Fant}(i)$	Set of received Fants of node i

During the route request phase, when a node receives a Fant, the operations are done according to Procedure recvFant. It first checks whether the Fant is in the received S_{Fant} set of node *i*. If the Fant is in S_{Fant}, it denotes that the Fant has arrived at node i, it does not do anything; If the Fant is not inS_{Fant}, it adds its node ID to Fant. Path, then it checks the DID to determine whether the entry is its own ID. If a node is the destination, it will create a Bant to the source node along the discovered path and initialize some parameters such as $E_{min}=E_{sum}=0$ and $HOP=0$. If the node is not the destination, it forwards Fant according to send Fant procedure continually.

```
  /* Node i receives a Fant*/
Procedure recvFant(i)
1. if Fant.ID is in SFant(i), then return
endif
2. it adds node i to Fant.Path and
Fant.ID to SFant(i)
3. If i is the destination d, then
4.     node i creates a Bant,
Bant.Esum(i) = 0,
5.     Bant.Emin(i)  =  0, Bant.HOP  =  0,
Bant.Path=Reverse(Fant.Path)
   else
6.      sendFant(i)
    endif
endProcedure
```

When a Fant reaches the destination *d*, node *d* creates a Bant as shown in Table 2. The task of the Bant is to return to the source node *s* along the path that was followed by the Fant and establishes the pheromone track to the destination node. Each node forwards Bant according to procedure sendBant. When a Bant arrives at the node *i* from its neighbor *j*, it updates corresponding parameters for path discovered by the Bant according to Procedure recvBant as follows:

$$HOP = HOP + 1, \quad E_{min} = min\{E_{min}, E_{cur(i)}\},$$
$$E_{sum} = E_{sum} + E_{cur(i)},$$
$$E_{avg} = \frac{E_{sum}}{(HOP+1)}, \qquad \phi_{id} = \frac{E_{avg} * E_{min}}{HOP+1}. \tag{1}$$

The pheromone of a node is the maximum pheromone of all paths at this node. When a Bant arrives at a node, it updates all parameters, the node forwards Bant to the next node in the path. When a Bant arrives at the source node, multi-paths have been built and the path discovery process is finished.

```
 /* Node i sends a Bant*/
Procedure sendBant(i)
1. If i is not the source, then
2.    it selects a next node j in
Bant.Path
3.    it sends Bant to j
   endif
endProcedure
 /* Node i receives a Bant from node j*/
Procedure recvBant(i)
1. update parameters of Bant
```

$\text{Bant}.HOP = \text{Bant}.HOP + 1$,

$\text{Bant}.E_{min} = \min\{\text{Bant}.E_{min}, E_{cur}(i)\}$,

$\text{Bant}.E_{sum} = \text{Bant}.E_{sum} + E_{cur}(i)$,

$E_{avg} = \frac{\text{Bant}.E_{sum}}{\text{Bant}.HOP}$, $\text{Bant}.\phi_{id} = \frac{E_{avg} * \text{Bant}.E_{min}}{\text{Bant}.HOP}$,

2. $\phi_{jd} = \max\{\text{Bant}.\phi_{jd}, \phi_{jd}\}$

```
3. If v is not the source, then
4.    sendBant(i)
   else
5.    start data transmission
   endif
endProcedure
```

2.3 Data transmission and route maintenance

When routes are discovered, the data packets can be sent through one of them. When a node *i* receives a data packet for a destination *d*, node *i* sends the data packet to a neighbor *j*, which is selected with probability $P_i(j)$. If *i* has no pheromone for the destination *d* in its pheromone table, *i* sends the data packet to a neighbor *j*, which is selected randomly. If node *i* has no neighbor, the data packet is discarded. In order to maintain the path and keep alive, the ACECR should update pheromone value dynamically. The traversal of each data packet increases the pheromone value of each link by $\phi_{id}=\phi_{id}*(1+\Delta_{id})$, where $\Delta_{id}=(E_{cur}(i))^{\beta}$ and β is a constant, generally we set β=0.1. To adapt dynamic network change in the ACO routing algorithm, each node evaporates a amount of pheromone at regular time intervals as $\phi_{id}=\phi_{id}*(1-\theta_{id}),\theta_{id}\in(0,1)$ is evaporation rate.

In order to explain the proposed ACECR protocol, an example network topology is shown in Fig. 1. There are 11 nodes in the network, each node has its energy. We assume that node 1 is the source and node 10 is the destination. When source 1 broadcasts a Fant packet to find the route paths, there are many return ants from destination 10, when Bants arrive at source 1, many paths are discovered with pheromone to the path listed at node 1 in Table 5. According to Table 5, a route table and route selection probability can be obtained by using Recv Bant Procedure and probability calculation formula as shown in Table 6. The multiple paths can be used to forward a data packet according to selected probability.

Node	Energy
1	8
2	7
3	6
4	2
5	8
6	2
7	5
8	5
9	2
10	7
11	9

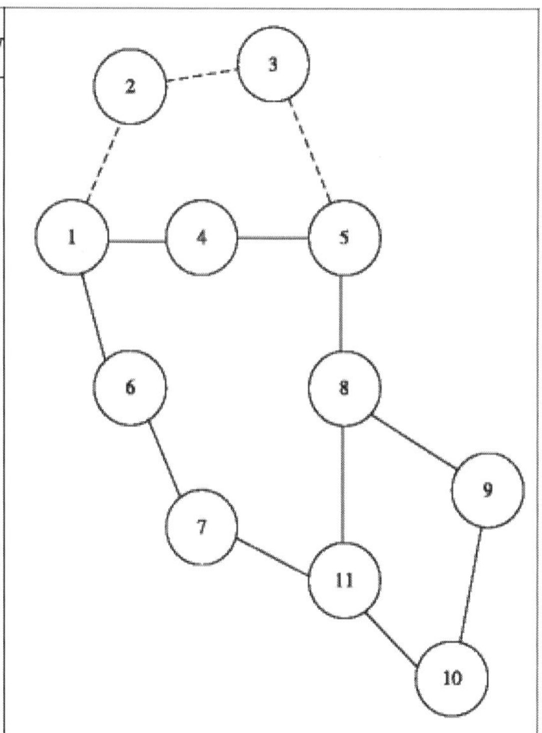

Figure 1. Network topology and energy distribution

Table 5. Discovered paths from 1 to 10 with pheromone at node 1

Paths	Pheromone
1-2-3-5-8-9-10	14.3
1-2-3-5-8-11-10	14.7
1-6-7-11-10	15.5
1-4-5-8-9-10	2.1
1-4-5-8-11-10	2.6

Table 6. Routing table and selection probability at node 1

Destination	Next hop	Pheromone	Probability
10	2	41.7	0.69
10	6	15.5	0.26
10	4	2.6	0.95

3. SIMULATION AND PERFORMANCE EVALUATION FOR DIFFERENT MOBILITY MODELS

In this section, we compare the performance of our proposed protocol ACECR to other two protocols AOMDA [3] and EAAR [6]. AOMDA protocol extends the single path AODV protocol to compute multiple paths, which always offers a superior overall routing performance than ADOV in a variety of mobility and traffic conditions. EAAR is an ACO-based energy-aware routing protocol, which not only incorporates the effect of power consumption in routing a packet, but also exploits the multi-path transmission properties of ant swarms and use min-max energy to calculate pheromone value; hence, it increases the battery life of a node. Mobility is a natural characteristic of ad hoc networks, it is imperative to use a mobility model that accurately represents the mobile nodes that will eventually utilize the given protocol. The choice of a mobility model can have a significant effect on the performance of an ad hoc network routing protocol. Many mobility models have been reviewed in [11], the mobility models are designed to describe movement pattern of mobile nodes including their locations, velocities and accelerations over time [12]. A mobility prediction-based routing protocol is proposed for DTNs in [13], which will be future research direction for ant conlony-based routing in MANETs. To the best of our knowledge, there is no research on ACO-based energy control routing protocols for different mobility models in ad hoc networks.

In this paper, we do not propose a new mobility model, our target is to measure the influence of the different mobility models to ant colony-based energy control protocols. We discuss following three different mobility models for three ant colony-based energy control protocols in ad hoc networks:

1) Random walk mobility model (Randwalk): each node moves from its existing location to a new location by randomly choosing an arbitrary direction and speed from a specified range [11].

2) Random waypoint mobility model (Randway): this model is equivalent to the random walk model except that the modification in speed and direction is done after predefined pause time [11].

3) Reference point group mobility (RPGM): this mobility model represents the random motion of a group of nodes as well as the random motion of each individual node within the group [14]. Group movements are based upon the path traveled by the logical center for the group.

3.1 Performance evaluation of routing protocols

NS-2 simulator is used to evaluate the performance of different protocols. There are 100 nodes in a network, which move over a square $1000m*1000m$ flat space. For RPGM model, we divided all nodes into four groups, there are 25 nodes in each group. Node's MAC layer uses IEEE-802.11 DCF media access control protocol, the radio transmission range and the interference range of nodes are all set to be 200 meter. Each node has a total energy of $100J$. Mobile nodes are assumed to move randomly according to the random walk, random waypoint, and RPGM mobility models. The speeds of nodes are set to be 1.5,5,10,15, and 20 per second, each node starts moving from a randomly selected initial position to a target position, which is also selected randomly in the simulation. Each packet size is 512-bytes, 10 constant-bit-rate (CBR) flows are generated randomly at a rate of 10 packets per second for 1000 s to test the performance of protocols.

In our simulation experiment, the following metrics are used for our performance study:

3.1.1 Data packet delivery ratio

The percentage of the number of data packets correctly delivered to the number of data packets sent by source nodes.

Figure 2 shows the packet delivery ratio of AOMDV, EAAR, and ACECR protocols at different speeds in different mobility models, where the packet delivery ratio for three routing protocols decreases when the speeds of nodes increase. We observe that the packet delivery ratio for ACECR and EAAR protocols is better than AOMDV protocol, because ACECR and EAAR protocols are energy control routing protocols, they can balance the energy use of the network, and reduce the link break caused by dead nodes. Since both average energy and the minimum energy of a path is considered in ACECR, it can select a path with more residual energy on global view, EAAR only considers the residual energy of nodes instead of paths, the packet delivery ratio for ACECR protocols is higher than that for AOMDV protocol.

3.1.2 Average end-to-end delay

The average time between transmission of data packets at sources and successful reception at receivers.

Figure 3 shows the average end-to-end delay of data packets from source nodes to their destination nodes for AOMDV, EAAR and ACECR in different mobility models, the end-to-end delays decrease with increase of node mobile speeds, because the increase of node mobile speeds will make network topology change, which will cause data buffer and route rediscovery. The average end-to-end delay for ACECR and EAAR protocols is less than AOMDV protocol, because ACECR and EAAR protocols are energy control routing protocols, and ant colony-based energy control routing protocol is multi-path routing protocols, they can balance the energy use of the network, and reduce the route rediscovery.

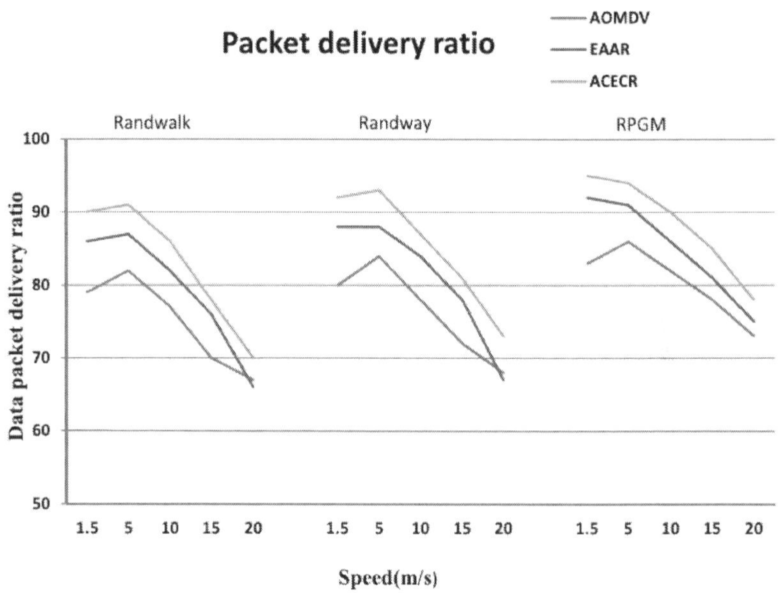

Figure 2. Packet delivery ratio for AOMDV, EAAR, and ACECR protocols in different mobility models

3.1.3 Routing load ratio

The percentage of the number of control packets sent and forwarded by all nodes to the number of all packets (control and data packets) propagated by nodes.

The communication overhead has profound impact on the performance of routing protocols, it represents the total size of exchanging packets in the network. The control packets increase the communication overhead and reduce the throughput of the network. Figure 4 shows that routing overhead of ACECR and EAAR protocols is higher than AOMDV protocol, since ACECR and EAAR protocols are multi-path routing protocols, they use pheromone updating to maintain the route selection.

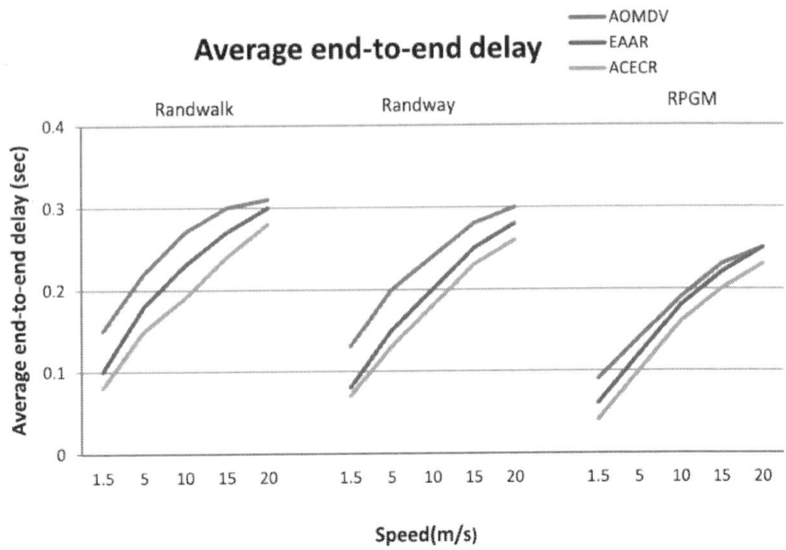

Figure 3. Average end-to-end delay for AOMDV, EAAR, and ACECR protocols in different mobility models

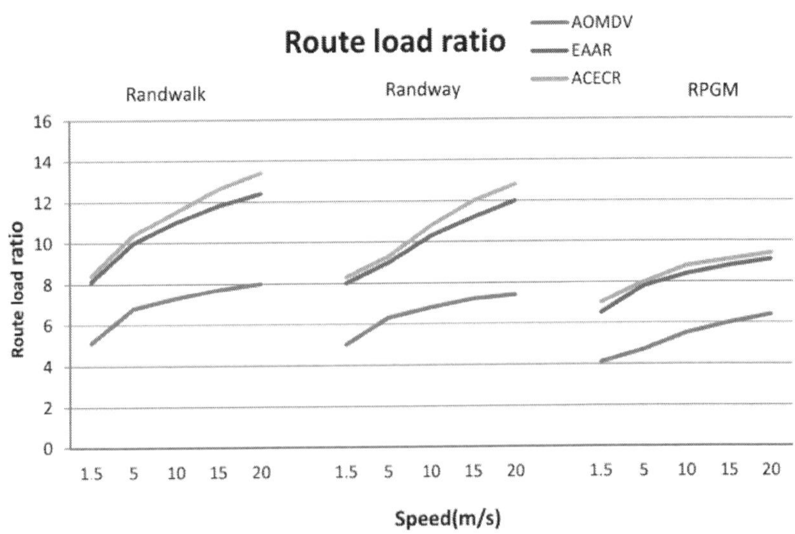

Figure 4. Route load for AOMDV, EAAR, and ACECR protocols in different mobility models

3.2 Energy consumption evaluation of nodes

Figure 5 shows the dead node ratio for AOMDV, EAAR, and ACECR protocols at different simulation times when nodes are moving at 15-m/s speed. In Fig. 5, the longer the simulation time is, the more there is dead nodes in the network, the ratio of dead nodes of ACECR and EAAR protocols is less than that of AOMDV protocol, since both average energy and minimum energy of a path is considered in ACECR, it can select a path with more residual energy on global view, EAAR only considers the residual energy of nodes instead of paths, AOMDV does not deal with energy balancing problem.

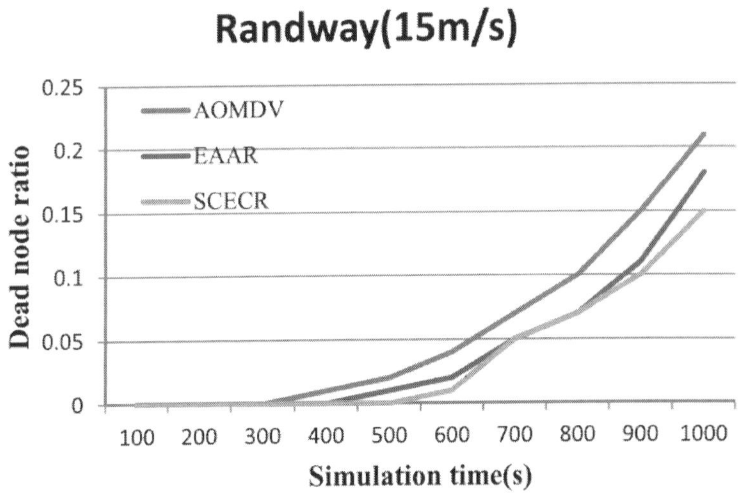

Figure 5. The node dead ratio for AOMDV, EAAR, and ACECR protocol at different times

Figure 6 shows the dead node ratio for ACECR protocols under in Randway, Randwalk, and RPGM mobility models at different simulation times when nodes are moving at 15-m/s speed. In Fig. 6, the more stable is the network topology, the less is the dead node ratio. If network topology is unstable, it will rediscover the routing paths and will consume more energy of nodes.

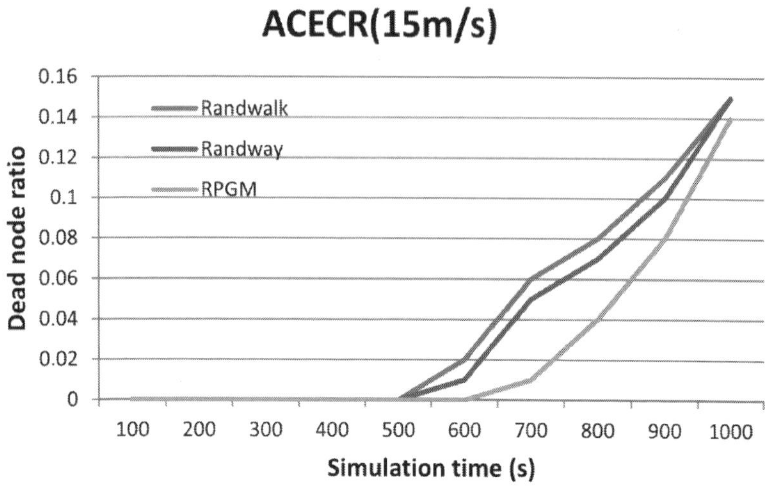

Figure 6. The node dead ratio for ACECR protocol in different mobility models

Figure 7 shows the relation between dead node ratio and node moving speed for ACECR protocols under Randway models. In Fig. 7, the higher is the speed of nodes, the more is the dead node ratio, this is because the increase of node moving speed will make network topology is unstable, it will rediscover the routing paths and will consume more energy of nodes.

4. CONCLUSIONS

In this paper, we propose an ant colony-based energy control routing protocol ACECR and evaluate the affect of different mobility models to the performance of ant colony-based energy control routing protocols in MANETs. In ACECR, the routing protocol will find the better route which has more energy than other routes through the analysis of average energy and the minimum energy of paths. Simulation results show that ACECR has a better performance than existing routing protocols, such as AOMDV and EAAR, in the number of dead nodes and the packet loss rate, which means that ACECR can extend the network lifetime. In addition, we test the performance of AOMDV, EAAR and ACECR for different mobility models in MANETs. Simulations investigate the movement characteristics of different mobility models and the effect on routing protocols. Simulation results show that ACECR has a better performance than the other two protocols in balanced energy consumption and extended network lifetime.

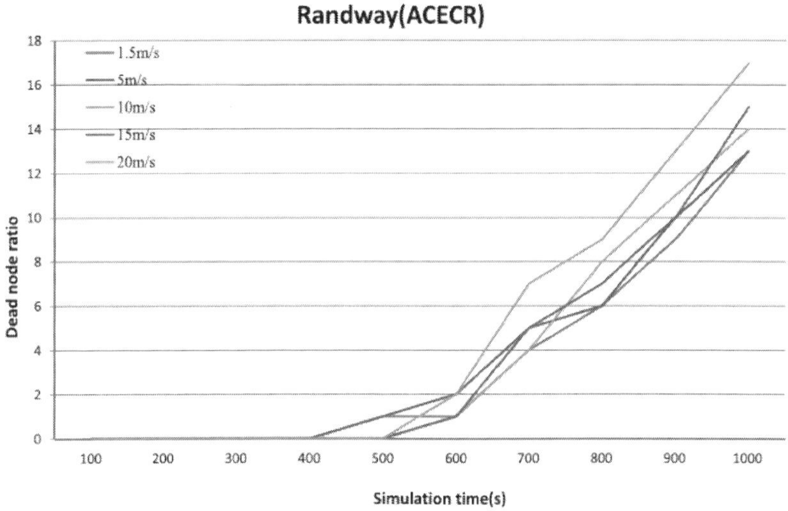

Figure 7. The node dead ratio nodes for ACECR protocol at different speeds

ACKNOWLEDGEMENTS

This work is supported in part by NSFC Grants 61373125, 61502201, 61272073, NSF-Guangdong Grants S2013020012865, 2014A0303101 72, and GDSTP Grant 2013B010401016

REFERENCES

1. M Dorigo, T Stuetzle, *Ant colony optimation*, (Bradford Company, Scituate, MA, 2004).

2. GD Caro, F Ducatelle, LM Gambardella, AntHocNet: An adaptive nature-inspired algorithm for routing in mobile ad hoc networks. Eur. Trans. Telecommun. **16**(5), 443–455 (2005).

3. MK Marina, SR Das, in *Proc. of Ninth International Conference on Network Protocols*. On-demend multi-path distance vector routing in ad hoc networks, (IEEERiverside, CA, 2001), pp. 14–23.

4. C Liang, AG Bourgeois, BH Yu, Power management in wireless ad hoc networks using AODV. SNPD/SAWN 2005. **5:**, 436–443 (2005).

5. B Shuang, ZB Li, JP Chen, An ant-based on-demand energy route protocol for IEEE 802.15.4 mesh network. Int. J. Wireless Inf. Netw. **16**(4), 225–236 (2009).

6. S Misra, SK Dhurandher, MS Obaidat, etc, An ant swarm-inspired energy-aware routing protocol for wireless ad hoc networks. J. Syst. Softw. **83:**, 2188–2199 (2010).

7. JP Zhou, JZ Lu, J Li, Ant-based balancing energy routing protocol for mobile ad hoc networks. J. Internet Technol. **14**(5), 835–842 (2013).

8. O Abedi, M Fathy, J Taghiloo, in *Proc. of IEEE/ACS International Conference on Computer Systems and applications, (AICCSA 2008)*. Enhancing AODV routing protocol using mobility parameters in VANET, (IEEEDoha, Qatar, 2008), pp. 229–235.

9. ZP Cai, ZZ Chen, GH Lin, in *Theor.Comput. Sci*, 410. 3.4713-approximation algorithm for the capacitated multicast tree routing problem, (2009), pp. 5415–5424.

10. ZP Cai, R Goebel, GH Lin, Size-constrained tree partitioning: approximating the multicast k-tree routing problem, Theor. Comput. Sci. **412**(3), 240–245 (2011).

11. AK Shukla, CK Jha, Mishra VK, in *Proc. of the Third International Conference on Soft Computing for Problem Solving, Advances in Intelligent Systems and Comoputing (SocPros 2014)*. Building of a competent mobility model for ad hoc wireless networks, (SpringerIndia, 2014), pp. 653–666.

12. Z Masoud, K Pejman, Using mobile node speed changes for movement direction change prediciton in a realistic category of mobility models. J. Netw. Comput. Appl. **36:**, 1078–1090 (2013).

13. LC Zhang, CP Cai, JL Lu, XM Wang, Mobility-aware routing in delay tolerant networks. Pers. Ubiquit. Comput. **19**(7), 1111–1123 (2015).

14. AH Jamalian, R Iraji, MT Manzuri-Shalmani, in *Proc. of 7th IEEE Int. Conf. on Cognitive Informatics (ICCI'08)*. A learing automata-based method for estimating the mobility model of nodes in mobile ad hoc networks, (IEEEStanford, CA, 2008), pp. 249–254.

2

PERFORMANCE EVALUATION OF VARIOUS ERASURES CODING TECHNIQUES IN DIGITAL COMMUNICATION

N. M. El-Gohary[1], M. A. M. El-Bendary[2], F. E. Abd El-Samie[3], M. M. Fouad[1]

[1]Department of Electonics and Communication, Faculty of Engineering, Zagazig University, Egypt

[2]Department of Communication Technology, Faculty of Industrial Education, Helwan University, Egypt

[3]Department of Electonics and Communication, Faculty of Electonics Engineering, Menofia University, Egypt

ABSTRACT

Reducing error in wireless, satellite, and space communication systems is critical. In the wireless communication system, various coding methods are employed on the data transferred to induce high bit error rates. High speed wireless networks, in order to address the large latency and degraded network throughput due to the retransmission triggered by frame loss, the purpose of this paper is to study and investigate the performance of fountain codes that is used to encode and decode the data stream in digital communication. This is an intelligent solution that encodes a number of redundant frames from the original frames upon link loss rate so that a receiver can effectively recover lost original frames without significant retransmissions. Since then, many digital Fountain coding methods have been invented such as Tornado codes, Luby transforms (LT) codes and Raptor codes.

KEYWORDS

Wireless communication systems, Fountain codes, Tornado codes, Luby transforms and Raptor codes

1. INTRODUCTION

In civil and military applications, wireless networking technologies have been widely deployed such as 3G/4G and IEEE 802.11 WLAN networks. However, wireless communication suffers from frame losses due to shadowing, mobility, channel fading, and interferences. Frame loss significantly undermines wireless network performance in that latency is enlarged and throughput is degraded. The large latency is induced by the retransmission of lost frames in the MAC layer [6].

On the Internet, data is transmitted in the form of packets. Each packet is armed with a header that describes the source and the destination of the packet, and often also a sequence number describing the absolute or relative position of the packet in a given stream. These packets are routed in the network from the source to the destination [16]. Due to various reasons, for example, buffer overflows at the intermediate routers; some packets may get lost and never reach their destination, other packets may be announced as lost if the internal check of the packet does not match. Therefore, the Internet is a very good real-world pattern of the BEC.

Reliable transmission of data over the Internet has been the subject of much research. Reliability is ensured by the use of suitable protocols. For example, the present TCP/IP ensures reliability by essentially retransmitting packets within a transmission window whose reception has not been acknowledged by the receiver. It is known that such protocols exhibit poor behavior in many cases, such as the transmission of data over heavily disabled channels or transmission of data from one server to multiple receivers or, such as poor wireless or satellite links. Moreover, ack-based protocols such as TCP perform poorly when the distance between the sender and the receiver is long, since large distances lead to idle times during which the sender waits for an acknowledgment and cannot send data. For these reasons, other transmission solutions have been proposed. One class of such solutions is based on coding. The original data is encoded using some linear erasure correcting code. If during the transmission some part of the data is lost, then it is possible to recover the lost data using erasure correcting algorithms. For applications it is critical that the codes used are able to correct as many erasures as possible, and it is also critical that the encoding and decoding algorithms for these codes are very fast. Reed–Solomon codes can be used to partially recover for the inefficiency of random codes. Reed–Solomon codes can be decoded from a block with the maximum possible number of erasures in time quadratic in the dimension. There are faster algorithms based on fast polynomial arithmetic, but these algorithms are often too complicated in practice. However, quadratic running times are still too large for many applications.

Other codes are constructed with linear time encoding and decoding algorithms that can come arbitrarily close to the capacity of the BEC [16, 18]. These codes, called Tornado codes, are very similar to Gallager's low-density parity-check (LDPC) codes [19], but they use a highly unequal weight distribution for the implied graphs.

In computer networks, fountain codes are ideally suited for transmitting information. A server sending data to many recipients can perform Fountain code for a given piece of data to generate an infinite stream of packets. As soon as a receiver requests data, the packets are copied and forwarded to the receiver. In a broadcast transmission model, there is no need for copying the data since any outgoing packet is received by all the receivers. In other types of networks, the copying can be done actively by the sender, or it can be done by the network. An example, if multicast is enabled; the recipient collects the output symbols, and leaves the transmission as soon as it has been received by them. That time, it uses the decoding algorithm to recover the original symbols. That, the number is the same regardless of the channel characteristics between the sender and the receiver. More loss of symbols just translates to a longer waiting time to receive the packets.

This paper surveys the channel coding methods in the physical layer and Digital Fountain code proposals in the application layer. The following paper is arranged as follows. Section 2 presents the RS codes architectures, applications and limitations. Section 3 discusses the fountain codes properties, the related erasure channel and construction. Section 4 Recent advances have produced powerful fountain codes, such as Tornado codes, Luby Transform (LT) codes and Raptor codes. Finally, section 5 discusses the advantages of fountain code over Reed Solomon code. Finally section 6 presents the conclusion.

2. DEVELOPING BACKGROUND RELATED WORK AND MOTIVATIONS

2.1. Error Correction Control (ECC)

In information and coding theory with applications in computer science and telecommunication system, error detection and correction are techniques that enables reliable and efficient delivery of digital data over communication channels. Many communication channels are subject to channel noise, and thus errors may be inserted during transmission from the source to a receiver. Error detection techniques permit detecting such errors, while error correction enables regeneration of the original data in many cases.

Error correction control is accomplished by adding redundant symbols to the message. These redundant symbols make it possible for the receiver to detect and /or correct some of the errors that may occur in the received message. The main challenge is to ensure the required protection against the determined transmission errors without paying too high a price in adding extra symbols.

Error correction may generally be accomplished in two different ways:

- *Forward error correction* (FEC): The sender encodes the data using an error-correcting code (ECC) before transmission. The receiver used the additional *redundancy* added the code to recover the original data.

- *Automatic repeat request* (ARQ): (referred also to as backward error correction): Every block of data received is checked using the error detection code used, and if the check fails, retransmission of the data is requested – this may be done repeatedly, until the data can be executed. This is an error control technique whereby an error detection scheme is combined with requests for retransmission of incorrect data.

FEC and ARQ may be combined, this is called *hybrid automatic repeat-request* (HARQ), such that minor errors are corrected without retransmission, and major errors are corrected via a request for retransmission. There are many different families of error-correcting codes of major importance for recording applications is the family of Reed-Solomon (RS) codes.

2.2. Binary Erasure Channel

The Binary Erasure Channel (BEC) is a channel model where the receiver either receives the transmitted bit or is informed with the erasure of the bit, that is, the bit was not received or erased. Therefore, the receiver has no idea about the transmitted bit with a certain probability p, and is exactly sure about the transmitted bit with a certain probability 1-p. According to Shannon, the capacity of BEC is 1-p, which means that for the alphabet size of 2^k, where k is the number of bits in the alphabet, no more than (1− p) k bits/symbol can be reliably communicated over the binary erasure channel [20].

Feedback from the receiver to the transmitter will not increase the capacity of the channel and reliable communication should be possible at this rate. Automatic Repeat Request (ARQ) schemes have so long been used as a classical approach to solve the reliable communication problem. However, redundant number of feedbacks used in the case of erasures causes improvident usage of bandwidth, network overloads and impossible delays. Also known as rateless erasure codes are a class of erasure codes with the property that a potentially limitless sequence of encoding symbols can be generated from a given set of source symbols such that the original source symbols can ideally be recovered from any subset of the encoding symbols of size equal to or only slightly larger than the number of source symbols. The term fountain or rateless refers to the fact that these codes do not exhibit a fixed code rate.

2.3. Reed-Solomon Code

Reed-Solomon (RS) code is a forward error-correcting code first invented in 1960 by Irving Reed and Solomon [1]. Only in the past few years has it become computationally possible to send high-bandwidth data using RS. Versions of Reed-Solomon codes are now used in error-correction systems found just about everywhere, including the storage devices, wireless communications, satellite communications and digital television.

RS codes are non-binary cyclic error-correcting codes. The RS encoder takes a block of digital data and adds extra information to the original data. While the errors occur during transmission or storage, the RS decoder processes, each block and attempts to correct errors and recover the original data. The number and type of errors that can be corrected depends on the characteristics of the RS code.

2.3.1. Encoding RS Codes

The basic structure of RS code is represented in Fig(1), as shown in that the codeword symbols (n) is unite of two segments information symbols (k) and parity symbols (2t). The information symbols (k) are having a message that is to be transmitted and parity symbols (2t) is the redundancy added to message to transmit it from source to destination without error [7].

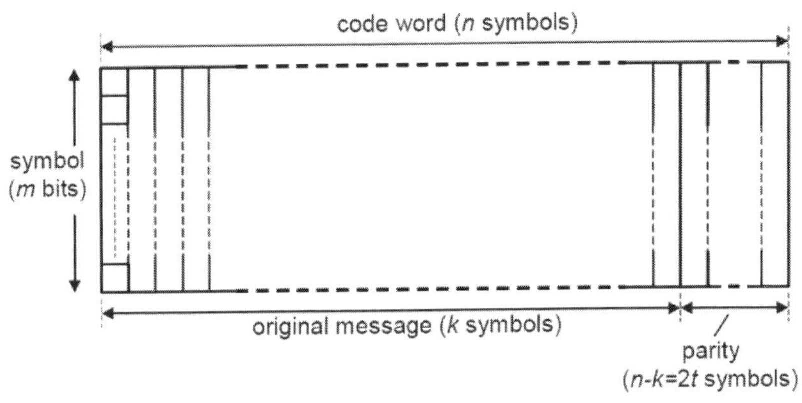

Figure 1. Encoding of RS codes

Reed-Solomon codes are nonbinary cyclic codes with symbols made up of *m*-bit sequences, where *m* is any positive integer having a value greater than 2. An RS code is specified as an RS (*n*, *k*) codes on *m*-bit

$$0 < k < n < 2^{m} + 2 \tag{1}$$

Where *k* is the number of data symbols being encoded, and *n* is the total number of code symbols in the encoded block. For the most conventional R-S (*n*, *k*) code,

$$(n, k) = (2^m - 1,\ 2^m - 1 - 2t) \tag{2}$$

Where t is the symbol-error correcting capability of the code, and $n - k = 2t$ is the number of parity symbols. An extended R-S code can be made up with or $n = 2^m$ or $n = 2^m + 1$.

Reed-Solomon codes attain the largest possible code minimum distance for any linear code with the same encoder input and output block lengths. For nonbinary codes, the distance between two code words is defined as the number of symbols in which the sequences differ. For Reed- Solomon codes, the code minimum distance is given by:

$$d_{min} = n - k + 1 \tag{3}$$

The code is capable of correcting any combination of t or fewer errors, where t can be expressed as [3]:

$$t = \left\lfloor \frac{d_{min} - 1}{2} \right\rfloor = \left\lfloor \frac{n-k}{2} \right\rfloor \tag{4}$$

RS differs from a Hamming code in that it encodes groups of bits instead of one bit at a time. Classical coding scheme for recovering erasures are Reed-Solomon codes [1, 3] employed in a variety of commercial applications, most notably in data storage as a key component of compact disks. In coding theory, Reed-Solomon codes are an example of Maximum Distance Separable (MDS) codes which achieve the Singleton bound [4]. The maximum distance separable (MDS) codes are practical codes that achieve the capacity of the erasure channel. A (*n, k, d*) MDS code, has a property that any *k* coordinates constitute an information set [2]. A receiver that receives any *k*symbols from a total of *n* symbols in each codeword can reconstruct the original message, provided it knows the position of the *k* received symbols. Reed Solomon (RS) codes are the most well-known MDS codes. These can be decoded in time *O* (*K2*), using algebraic methods such as list decoding.

2.3.2. Decoding of RS Codes

The RS decoder consists of two main stages; (1) error detection stage, and (2) error correction stage as shown in Fig. (2) [15]. Firstly, a serial syndrome is used to check if this codeword is a valid codeword or not. If errors occurred during transmission, the decoder carried out error detection, then try to correct these errors. Secondly, the key equation solver is used as decoding algorithm to find the coefficients of the error-location polynomial σ(x) and error-evaluator polynomial W(x). Thirdly, the Chien search block which is used to find the roots of σ(x) which present the inverse of the error locations. Fourthly, the Forney algorithm block is used to find the values of the errors. Finally, after getting the values and locations of the error, the received code word can be corrected by XOR-ing the received vector with the error vector.

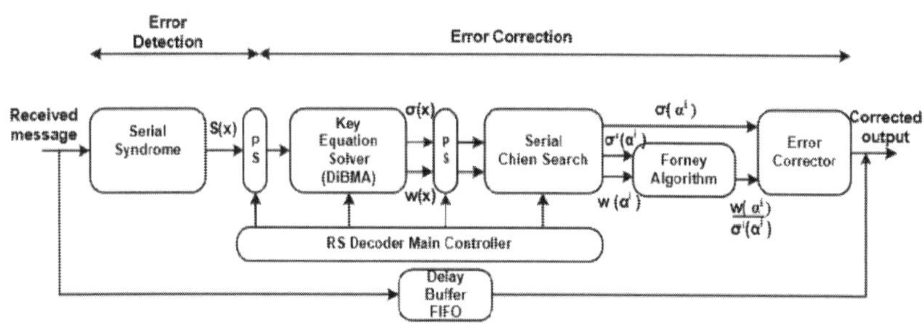

Figure 2. RS decoder

2.3.3. Reed-Solomon Code Applications

Reed-Solomon codes are the most frequently used digital error control codes in the world, due their usage in computer memory and non-volatile memory applications. A hurried list of significant applications includes the Digital Audio Disk, Deep Space Telecommunication Systems, and Error Control for Systems with Feedback, Spread-Spectrum Systems, and Computer Memory [8].

2.3.4. Reed-Solomon Codes Limitations

RS codes are not proper for bulk data distribution over the internet although spreading. When data rates are of the order of Mbps, the quadratic decoding time is unacceptable [2]. Moreover, typical RS code implementations have small block lengths such as the NASA standard (255; 233; 33) code. This requires a large file to be divided into many small blocks before transmission. Finally, since RS codes are block codes, they need to be designed for a specific rate. This requires that we need previous to evaluate the erasure probability of the channel. It is not possible when multiple clients over different quality of channels are being served simultaneously.

3. FOUNTAIN CODES

A fountain code is a forward-error-control code that can produce as many redundant packets as needed for packet erasure correction. Unlike automatic-repeat-request (ARQ) transmission, fountain coding does not require the destination to inform the source of the identities of the packets that are erased or even keep track of which packets are erased. We discuss the use of fountain coding for both unicast and multicast transmission in packet radio systems, where communication occurs over time-varying channels with fading, shadowing, and other types of propagation losses.

A decoding algorithm for a Fountain code is an algorithm which can recover the original k input symbols from any set of n output symbols with high probability [16]. For good Fountain codes the value of n is very close to k, Note that the number n is the same regardless of the channel characteristics between the sender and the receiver. More loss of symbols just translates to a longer waiting time to receive the n packets. For noisy wireless channels, the waiting time and then the overall throughput performance of the fountain coding system depend heavily on the selection of the channel code and the modulation format used to transmit the wireless signals.

Proposed for wireless mesh networks by Katti, et al not only forwards the packets, but also mixes packets from different sources into a single transmission and resolves the packets at the receiver. In upper layers, a coding concept called Digital Fountain has been introduced in 1998 by Byers, et al to generate a stream of packets including some redundant packets, like in a water fountain to address the potential packet loss in multicast applications that do not allow retransmission. Since then, many Digital Fountain coding methods have been invented such as Luby transform coding and Raptor codes.

Consider a setting where a large file is disseminated to a wide audience who may want to access it at various times and have transmission links of different quality. Current networks use unicast established protocols such as the transport control protocol (TCP), which requires a transmitter to continually send the same packet until acknowledged by the receiver. It can easily be seen that this architecture does not scale well when many users access a server concurrently and is extremely inefficient when the information transmitted is always the same. In fact, TCP and other unicast protocols place strong importance on the ordering of packets to simplify coding at the expense of increased traffic.

3.1. Digital Fountain Codes

The digital fountain was devised as the ideal protocol for transmission of a single file to many users who may have different access times and channel fidelity. The name is drawn from an analogy to water fountains, where many can fill their cups with water at any time. The output Packets of digital fountains must be universal like drops of water and hence be useful independent of time or the state of a user's channel [12, 13].

The encoder of fountain codes is like a fountain spewing as show in Fig (3). Infinite coded symbols can be produced. Source data is divided into *k* input symbols of size *l*. With fountain codes, the *k* input symbols are combined into infinite encoding symbols at source. All *k* input symbols can be cured from any set of $(1+\varepsilon)$ *k*encoding symbols, where $0< \varepsilon <1$. Encoder of fountain codes is bit rate independent which is not limited by the size of the source data and can generate an unlimited number of encoding symbols.

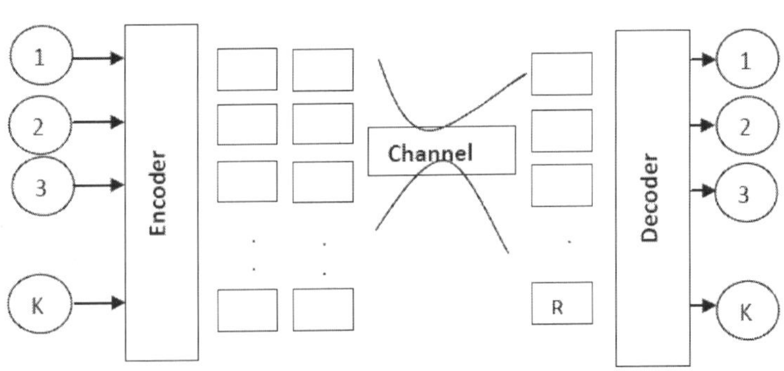

Figure 3. Fountain code

3.2. Fountain Codes Properties

Consider a file that can be split into k packets or information symbols and must be encoded for a BEC. A digital fountain that transmits this file has the next properties:

1) It can generate an endless supply of encoding packets with constant encoding cost per packet in terms of time or arithmetic operations.

2) A user can reconstruct the file using any k packets with constant decoding cost per packet, meaning the decoding is linear in k.

3) The space needed to store any data during encoding and decoding is linear in k.

These properties show that the digital fountains are as reliable and efficient as TCP systems, but also universal and tolerant, properties desired in networks.

3.3. Fountain Code Construction Outline

Fountain Codes are a new class of codes designed and ideally suitable for dependable transmission of data through an erasure channel with unknown erasure probability. Infinite number of output symbols can be produced potentially by the encoder. Output symbols can be bits or more general bit sequences. However, random linear Fountain Codes have encoding complexity of $O(N^2)$ and decoding complexity of $O(N^3)$ which makes them impractical for nowadays applications.

The fountain code constructions we provide all have the property that encoded symbols are generated separately from one another. Moreover, we will assume that the set of receiving encoded symbols is independent of the values of the encoded symbols in that set, an assumption that is often true in practice. For a given value of *k*, these assumptions mean that the probability of the inability of the decoding is only dependent on how many encoded symbols are received and independent of the pattern of which encoded symbols are received.

3.3.1. Fountain Coding

If the original k source symbols can be recovered from any k encoding symbols, fountain code is maximized. Fountain codes are known that have efficient encoding and decoding algorithms and that permit to pick up the original k source symbols from any k' of the encoding symbols with high probability, where k' is just slightly larger than k.

Digital fountains have changed the standard transmission model. A digital fountain can encode and transmit an infinite number of data packets until every user has enough information to ensure correct decoding. Emerging peer-to-peer applications in multimedia broadcasting are only two examples of many other scripts where digital fountains can be applied successfully.

Consider a file that can be split into k packets or information symbols and must be encoded for a BEC as shown in Fig (4) [17]. Regardless of the erasure probability, Fountain Codes are near optimal for all BEC. Therefore, on the BEC, Fountain Codes are called universal codes. A message consists of $k*k$ bits and each drop contains k bits.

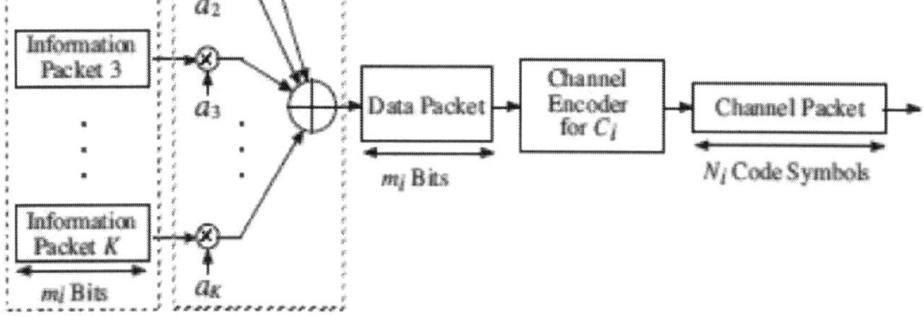

Figure 4. Fountain code

Whoever collects any $K' \geq K$ number of k bits, where K' is little larger than K, can recover the original message with high probability.

Fountain Codes can be accomplished at random with an average degree of k / 2. The degree is the number of one's divided by the total number of bits in the generator matrix. The average degree of the generator matrix determines the complexity of encoding and decoding process. The higher the degree, the higher the complexity at the transmitter and receiver side and the more successful the receiver is in the decoding phase. Let us assume that we transmit k source symbols k s, s, s... s 1 2 3 with a random generator matrix of degree k / 2. The encoding process of Fountain Codes is given by the following equation [9]:

$$t_n = \sum_{k=1}^{k} s_k G_{kn} \tag{5}$$

Where t_n indicates to the transmitted symbols. G_{kn} Can be generated at the transmitter side pseudo-randomly with a random seed, namely by a key, and transmitted to the receiver causing an extra overhead cost. Therefore, the symbol size is much larger than the key size, this overhead is neglectable. One other way to produce a unique G_{nk} is to synchronize the receiver and the transmitter with same clock pulses and to use deterministic random number generators on both sides.

3.3.2. Fountain Decoding

The decoding process of Fountain Codes is given as:

$$s_k = \sum_{n=1}^{k} t_n G_{nk}^{-1} . \tag{6}$$

In order for a $k*k$ G matrix to be invertible, each row should be linearly independent from the others. The probability that the first row is not an all zero row is $1\text{-}2^{-k}$, the probability that the second row is neither all zero nor same with the first row is $1 - 2^{-k+i}$. Iterating until K, we get as the overall success rate:

$$1-\delta = \prod_{i=0}^{k-1} (1 - 2^{-k+i}) \tag{7}$$

1−δ is lower bounded by 0.289 for $k > 10$. For any NxK binary matrix to be invertible, δ is upper bounded by $2^{-(N-K)}$. Accordingly, each additional row increases the success probability drastically. Thus, as the message size increases, random Fountain Codes come arbitrarily close to the channel capacity. Despite a very small overhead and rate erasure independency, random Fountain Codes have a quadratic encoding complexity, k bits times the degree $k/2$, and cubic decoding complexity $\sim 2K^3/3$. This makes them far away from most of the applications such as mobile broadcasting, where only a limited processor power can be used on the receiver side.

Then, many digital Fountain coding methods have been invented such as Tornado codes, Luby transform codes and Raptor codes.

4. FOUNTAIN CODING METHODS

4.1. Tornado Code

4.1.1. Introduction

Tornado codes first appeared in a technical report in 1997. Tornado codes are a new class of erasure randomized codes which have linear-time encoding and decoding algorithms. They can be used to transmit over erasure channels at rates extremely close to capacity. For Tornado codes, the encoding and decoding algorithms are both simple and faster by orders of magnitude than the best software implementations of standard erasure codes. Tornado codes will be expected extremely useful for applications such as reliable distributions of bulk data, including video distribution, software distribution, news and financial distribution, popular web site access, and military communications.

We consider a system model in which a single transmitter performs evaluated data transfer to a larger number of users on an erasure channel. Our objective is to achieve complete file transfer with the minimum number of encoding symbols and low decoding complexity. For k information symbols, RS codes can achieve this with k log k encoding and quadratic decoding times. The reason for the longer decoding time is that on RS codes, every redundant symbol depends on all information symbols. In Tornado codes by contrast, every redundant symbol depends only on a small number of information symbols. Thus they achieve linear encoding and decoding complexity, with the cost that the user requires slightly more than k packets to successfully decode the transmitted symbols. The main contribution is the design and analysis of optimal degree distributions for the bipartite graph such that the receiver is able to recover all missing bits by a simple erasure decoding algorithm. The innovation of Tornado code has also inspired work on irregular LDPC codes.

4.1.2. Construction

The design and analysis of Tornado codes are mathematically simple and interesting. The design requires careful choice of a random irregular bipartite graph where the structure of the irregular graph is very important. The progress of the decoding algorithm is designed by a simple AND-OR tree analysis, which instantly gives rise to a polynomial in one variable with coefficients determined by the graph structure. We design a graph structure based on these polynomials assure successful decoding with high probability.

Tornado codes are erasure block codes based on the irregular spare graph. Given an erasure channel with loss probability p, they can correct up to p (1- ε) errors. They can be encoded and decoded in time proportional to n^* log(1/ ε). There are eight input symbols named $x1$, $x2$... $x8$ as shown in Fig (5). With tornado codes, four encoding symbols named $y1$, $y2$, $y3$ and $y4$ is produced by eight input symbols. Tornado codes can support that any one of$y1$, $y2$, $y3$ and $y4$ can be recovered by three others.

Tornado codes are not suitable for large data transfer systems because the complexity of encoding and decoding algorithms for tornado codes is proportional to block length. For encoding the large amounts of data, an alternative classical ARQ concept is used as on the Internet.

4.2. LT Code

4.2.1. Introduction

LT codes are the first implementation of digital fountain codes in 2002 proposed by Michael Luby. Luby Transform (LT) codes have been invented to reduce the encoding and decoding complexity of random linear Fountain Codes while the small overhead is ensured. With a good choice of degree distributions of the edges in the Tanner graph, LT codes randomly close to channel capacity with certain decoder reliability and logarithmically inducing increases in encoding and decoding costs. In LT codes, Data was divided into blocks with fixed size. Each block is divided into fixed size symbols. So the number of input symbols is fixed. The encoder of LT codes can generate an infinite number of encoded symbols. When the number of encoding symbols is received slightly larger than the number of input symbols, this means the decoder of LT codes can recover all input symbols.

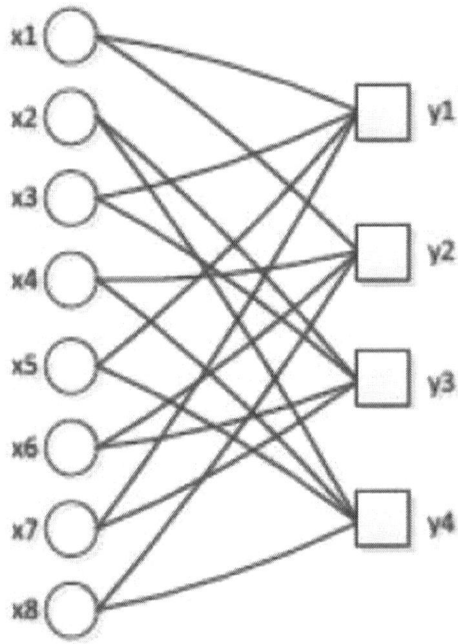

Figure 5. Tornado codes

Reducing the degree distribution resulting linear time encoding and decoding complexity leads to decrease in the reliability of the decoder. Thus, the decoder cannot decode all the input symbols with the lower degree distribution for the same overhead limitation. So, correcting the erasures arising from the weakened decoder would utilize an erasure correcting pre-code.

4.2.2. The Construction

For the binary erasure channel, LT codes are considered the first practical rateless codes. The encoder can produce as many encoding symbols as required to decode k information symbols. For LT codes, the encoding and decoding algorithms are simple and similar to parity-check processes. LT codes are efficient in that the transmitter does not require an acknowledgement (ACK) from the receiver. This property is especially desired in multicast channels because it will significantly decrease the overhead incurred by processing the ACKs from multiple receivers.

LT codes are known to be efficient if k information symbols can be recovered from any $k + O(\sqrt{k}\, ln^2\,(k/\delta))$ encoding symbols with probability $1 - \delta$ using $O\,(k.\ln(k/\delta))$ operations. However, their bit error rates cannot be decreased below some lower bound, meaning they suffer an error floor.

To reduce the computational complexity, the number of edges at the encoder side should be reduced. LT codes with simple encoding and decoding algorithms can then be considered as sparse random linear Fountain Codes. Although, there are simple and fixed encoding and decoding schemes defined for LT codes the degree distributions of the edges play a crucial role in the design of good codes. Good codes are such codes, which have low encoding and decoding costs as well as a small overhead and a decoding failure. Let us start with the definitions of encoding and decoding schemes.

A. LT Coding

Any number of encoding symbols can be independently generated from k information symbols by the following encoding process:

1) Determine the degree d of an encoding symbol. The degree is chosen at random from a given node degree distribution P (x).

2) Choose d distinct information symbols uniformly at random. They will be neighbors of the encoding symbol.

3) Assign the XOR of the chosen d information symbols to the encoding symbol. This process is similar to generating parity bits except that only the parity bits are transmitted.

The degree distribution P (x) in Fig(6), comes from the sense that we can draw a bipartite graph, in which consists of information symbols as variable nodes and encoding symbols as factor nodes. The degree distribution determines the performance of LT codes, such as the number of encoding symbols and probability of successful decoding. The degree distribution is analyzed.

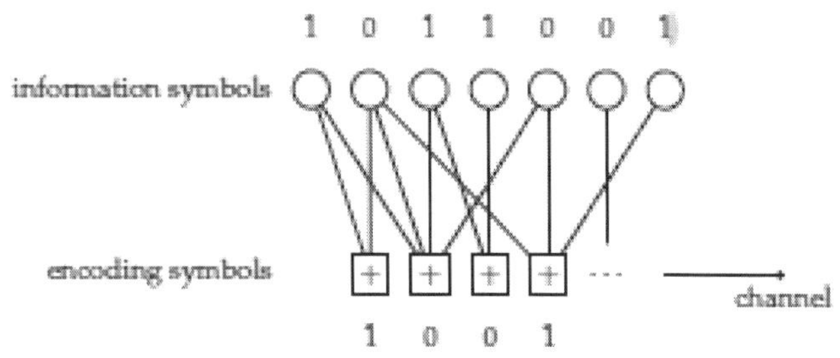

Figure 6. Generation of encoding symbols

The encoding symbols are transmitted through a BEC with the probability of erasure p. The special characteristic of a BEC is that receivers have correct data or no data. There is no confusion where the decoder needs to "guess" the original data; it recovers the true data or gives up.

B. LT Decoding

For decoding of LT codes, a decoder needs to know the neighbors of each encoding symbol. This information can be transferred in several ways. For example, a transmitter can send a packet, which consists of an encoding symbol and the list of its neighbors. There is an alternative method is the encoder and the decoder share a random number generator seed, and the decoder finds out the neighbors of each encoding symbol by generating random linear combinations synchronized with the encoder.

4.2.3. Degree Distribution

LT codes do not have a fixed rate and hence the desired property is that the probability of successful recovery is as high as possible while the number of encoding symbols required is kept small. Describing the property in the terminology of the LT process:

- For encoding symbols, the release rate is low in order to keep the size of the ripple small and prevent waste of encoding symbols.
- For encoding symbols, the release rate is high enough to keep the ripple from dying out.

Therefore, the degree distribution of encoding symbols needs to be very well designed so as to balance between the trade-off. This is the reason that the degree distribution plays an important role in LT codes. Several probability distributions are investigated as the degree distribution used in the fame Fountain encoding process [10] as follows:

1) Uniform distribution:

$$p_{i=1/n} \qquad \forall_i = 1,2,3, \text{------------}, n \tag{8}$$

2) Normal distribution:

$$\mu = [\, n/2], \qquad \delta = k/2 \tag{9}$$

$$p_i = \frac{1}{\sqrt{2\pi\delta^2}}\, e^{-\frac{(x_1-\mu)^2}{2\delta^2}} \qquad \forall_i = 1, \ldots\ldots, k; \tag{10}$$

Where $x_i = [\text{randn}*\delta + \mu]$

3) Sequential distribution:

$$p_i = \frac{1}{n-[n/k]} \qquad \forall_i = \left[\frac{n}{k}\right], \ldots., n \tag{11}$$

4) Ideal solution distribution

$$p_1 = 1/n \tag{12}$$

$$p_i = 1/i(i-1) \quad \text{For } i = 2, 3, \text{----}, n \tag{13}$$

5) Robust Soliton distribution, $R = c \ln\left(\frac{n}{\delta}\right)\sqrt{n}$ where c and δ are extra parameters; c > 0 is some suitable constant.

$$\tau_i = \begin{cases} \frac{R}{i_n}, & for\ i = 1,2,\ldots\ldots\ldots\left(\frac{n}{R}\right). \\ \frac{R}{n} ln\left(\frac{R}{\delta}\right). & for\ i = \left(\frac{n}{R}\right).\ o \\ 0, & otherwise. \end{cases} \quad (14)$$

Enough frames can be encoded together as redundant frames to make sure there are enough variety of encoded frames at the receiver.

4.3. Raptor Code

4.3.1. Introduction

Raptor code has been standardized in the 3GPP (Third Generation Partnership Project). Raptor Codes are an extension of LT codes combined with a pre-coding scheme. Raptor (rapid Tornado) codes were developed and patented in 2001 as a way to reduce decoding cost to O (1) by preprocessing the LT code with a standard erasure block code (as a Tornado code). The main idea of Raptor Codes is to refresh the condition of recovering all input symbols and to require that only a constant fraction of input symbols be recoverable. Then the number of edges in the Tanner graph will exhibit only a constant degree which will precede linear time encoding and decoding costs. This is done by utilizing an erasure correcting pre-code working in linear time. The degree distribution, which is used for Raptor Codes, should be completely different from the one that of LT codes. Because, in the concept of Raptor Codes, we are forced to recover as many input symbols as possible for a given constant average degree rather than to recover all input symbols to be recovered and decoding to be successful. Degree distribution design and pre-coding is the heart of Raptor Codes. The degree distribution, which is used for Raptor Codes, should be completely different from the one that of LT codes. From the concept of Raptor Codes, we are forced to recover as many input symbols as possible for a given constant average degree rather than to recover all input symbols while maintaining the small overhead.

4.3.2. The Construction

Digital Fountain, Inc. proposed Raptor codes in 2006. It is a concatenation of a systematic pre-code with LT codes. As shown in Fig (7), in the pre-code, *k* native symbols are first mapped to (1+ε) *k* pre-coded symbols. Infinite coded symbols can be generated from pre-coded symbols by LT codes. In decoding process of Raptor codes, pre-coded symbols are recovered by LT codes firstly, and then input symbols are recovered by pre-coded symbols.

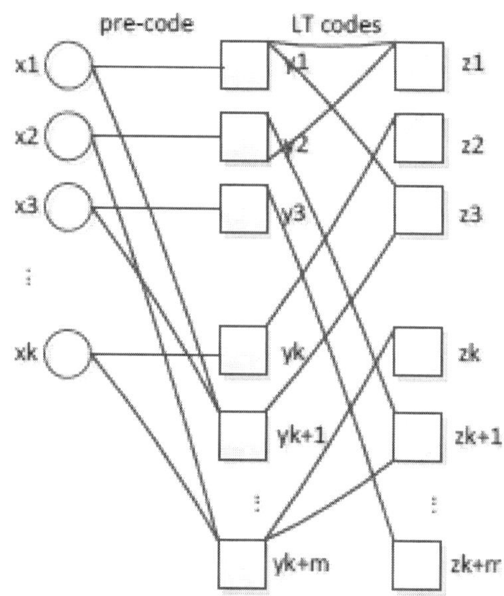

Figure 7. Raptor codes

A Raptor code can achieve constant per-symbol encoding and decoding cost with overhead close to zero and a space proportional to k. This has been shown to be the closest code to the ideal universal digital fountain. A similar vein of work was proposed in under the name online codes.

We have already seen two extreme cases of Raptor codes. When there is no pre-code, then we have the LT code. As an example of a pre-code only (PCO) Raptor code for the s are an extension of LT codes [11].

Raptor Codes can produce a potentially infinite stream of symbols such that any subset of symbols of size k (1+ε) is sufficient to recover the original *k* symbols with high probability. The original symbols are recovered from the collected ones with *O* (k log (1/ε)) operations and output symbol is generated using *O* (log (1/ε)) operations. Raptor encoding starts with a suitable design of the pre-code. Shokrollahi uses LDPC codes as a pre-code with a constant rate of (1+ε /2) (1+ε) and BP algorithm can work in linear time and decode (ε / 4) (1+ε) fraction of erasures where ε is a real positive number. Next, the intermediate symbols are encoded with LT coding using a suitable degree distribution.

Raptor decoding starts with the LT decoding process. In the example, LT decoding can recover all the intermediate symbols, but the ones filled with black. Since the pre-code is systematic, the first three input symbols are immediately recovered. The fifth intermediate symbol is encoded by xor-ing the third and fourth input symbol. We can recover the fourth input symbol by adding the fifth intermediate symbol to the third input symbol. Hence, as it is seen that the decoding process succeeds. Decoding is done the same way as described in this paper for LT decoding. LDPC decoding is performed using the BP algorithm.

The table below summarizes the characteristics of the various codes that are designed for the digital fountain ideal:

Table 1. Summary of fountain codes

	Tornado	LT	Raptor
Rateless	No	Yes	Yes
Overhead	ϵ	$\epsilon \to 0$	$\epsilon \to 0$
Encoding complexity per symbol	$O(\epsilon \ln(1/\epsilon))$	$O(\ln(k))$	$O(1)$
Decoding complexity per symbol	$O(\epsilon \ln(1/\epsilon))$	$O(\ln(k))$	$O(1)$
Space per symbol	$O(1)$	$O(1)$	$O(1)$, with a larger constant.

5. ADVANTAGES OF FOUNTAIN CODE OVER REED SOLOMON CODE

- A DF Raptor code with *k* input symbols that produces *n* output symbols using symbols of size *S* bytes is designated Raptor

(n,k,S).The Digital fountain code does not have any limitation to the amount of data that can be protected: the number of source symbols k for a DF Raptor code can be as large as desired, and the symbol size S is not subject to any constraints other than that it be less than or equal to the packet size. As a result, the value of S can be much smaller for a DF Raptor code than is possible for a Reed-Solomon erasure code.

- A DF Raptor code's processing requirements are significantly less than that of a Reed-Solomon erasure code and grow only linearly with the source block size, while a Reed-Solomon erasure code's processing requirements grow quadratic ally with source block size.

- In streaming applications, the erasure correction performance of Reed-Solomon erasure codes is directly affected by the inefficiencies associated with the limited number of symbols that can be used by the Reed-Solomon algorithm, while Reed-Solomon erasure codes can require an order of magnitude greater processing power for encoding and decoding than a DF Raptor code.

- The Digital Fountain code's offer excellent performance compared to the Reed-Solomon erasure codes. For a given rate of packet loss, the mean time between artifacts for the DF Raptor code is significantly greater than that of the Reed-Solomon codes As can be seen from Fig (8).

- In streaming applications, the length of the protection period reflects a trade-off between robustness and latency. In general, the longer the protection period, the more likely that the actual number of lost packets over that period will approach the average packet loss rate so that a fixed code rate erasure correction code can be sure to provide sufficient protection. The DF Raptor code provides greater protection against packet loss, one can arbitrarily choose any protection period and latency without constraint as shown in Fig (9). The Reed-Solomon erasure code with the same amount of overhead and with the same mean time between artifacts, the protection period that can be supported by a single Reed-Solomon code is limited. With Reed-Solomon codes, however, a protection period of data must be segmented into multiple blocks that are protected individually, and the resulting encoded blocks are then interleaved, at the cost of additional processing as well as reduced packet loss protection.

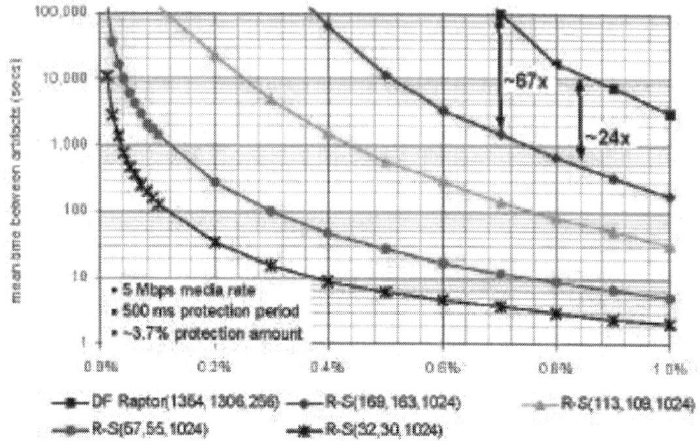

Figure 8. A DF Raptor code can provide a mean time than that provided by Reed-Solomon erasure codes [21]

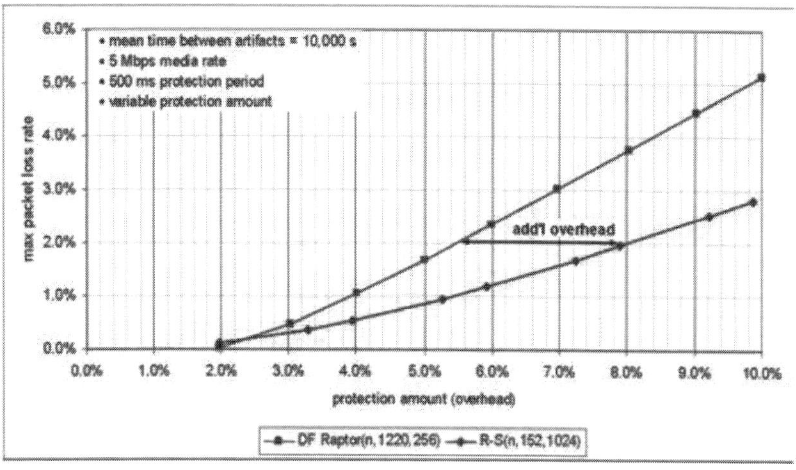

Figure 9. The maximum packet loss rate for the DF Raptor code is higher than for the Reed-Solomon code [21]

6. THE CONCLUSIONS

It is clear, that the Fountain codes have flexible applications at a fixed code rate cannot be determined previously, and where efficient encoding and decoding of large amounts of data is required as:

1) When multiple clients over different quality of channels are being served simultaneously.

2) The decoding time is acceptable when data rates are of the order of Mbps.

3) Fountain codes are suitable for large data distribution over the internet.

4) Raptor Q code, is the more advanced Raptor code with greater flexibility and improved reception overhead. This code has been introduced into the IETF. It can be used with up to 56,403 source symbols in a source block, and a total of up to 16,777,216 encoded symbols produced for a source block. This code is able to recover a source block from any collection of encoded symbols equal to the number of source symbols in the source block with high probability and seldom cases from slightly more than the number of source symbols in the source block.

REFERENCES

1. S. Reed and G. Solomon "Reed-Solomon code". Polynomial codes over certain _nite _elds. Journal of the Society for Industrial and Applied Mathematics, 1960.

2. David Forney Jr., Unpublished Course notes., http://ssg.mit.edu/6.451, 2003.

3. S. Wicker and V. Bhargava, Eds., "Reed-Solomon Codes and Their Applications" IEEE Press, Piscataway, NJ, USA, 1994.

4. J. van Lint, "Introduction to Coding Theory" Springer-Verlag (2nd Ed.), 1992.

5. G. Joshi, J .Bum Rhim, John Sun, Da Wang, "Fountain codes", December 7, 2010.

6. Qingmei Yao, Chong Tang, Shaoen Wu Frame Fountain: Coding and Decoding MAC Frames.

7. International Journal of Advances in Science and Technology, BER Performance of Reed-Solomon Code Using M-ary FSK Modulation in AWGN Channel Vol. 3, No.1, 2011.

8. V. Korrapati, M V D Prasad*2, D. Venkatesh, G. Arun Tej "A Study on performance evaluation of Reed Solomon Codes through an AWGN Channel model for an efficient Communication System" International Journal of Engineering Trends and Technology (IJETT) - Volume4Issue4- April 2013.

9. G. GÜL, S. Adhikari and E. Mustafin "Fountain Codes, LT Codes, and Raptor Codes".

10. Q. Yao, C. Tang, Sh. Wu School of Computing University of Southern Mississippi, USA "Frame Fountain: Coding and Decoding MAC Frames".

11. Shokrollahi and M. Luby Foundations and Trends in Communications and Information Theory "Raptor Codes" Vol. 6, Nos. 3–4 (2009) 213–322 _c 2011.

12. J. Byers, M. Luby, M. Mitzenmacher, and A. Rege, "A Digital Fountain Approach to Reliable Distribution of Bulk Data," In Proceeding of SIGCOMM, Vancouver, BC, CA, September 1998.

13. J. Liao, L. Zhang, X. Zhu, J. Wang and M. Liao " Loss Rate Based Fountain Codes for Data Transfer".

14. Khisti "Tornado Codes and Luby Transform Codes" October 22, 2003.

15. H. A. A. Elsaid, "Design and implementation of Reed-Solomon decoder using decomposed inversion less Berlekamp-Massey algorithm", M. Sc. Thesis, Faculty of Engineering, Cairo University, 2010.

16. Shokrollahi, \Raptor codes," IEEE Transactions on Information Theory, vol.52, no. 6, pp. 2551{2567, June 2006.

17. J. D. Ellis, "Adaptive Transmission Protocols For Wirless Communication Systems With Fountain Coding" A Dissertation Presented to the raduate School of Clemson University, December 2013.

18. M. Luby, M. Mitzenmacher, A. Shokrollahi, and D. Spielman, "Efficient erasure correcting codes," IEEE Trans.Inf. Theory, vol. 47, no. 2, pp. 569–584, Feb. 2001.

19. R. G. Gallager, "Low Density Parity-Check Codes". Cambridge, MA: MIT Press, 1963.

20. D. J.C. MacKay Cavendish Laboratory, University of Cambridge" Fountain codes" 1998 IEE, Cagliari, Italy Proc. of the Fourth Workshop on Discrete Event Systems.

21. San Diego "Why DF Raptor is Better Than Reed-Solomon for Streaming Applications", U.S. Dept. of Commerce, 2010.

3

WIRELESS SENSOR NETWORK NODE OPTIMAL COVERAGE BASED ON IMPROVED GENETIC ALGORITHM AND BINARY ANT COLONY ALGORITHM

Jingwen Tian*, Meijuan Gao and Guangshuang Ge

College of Information Technology, Beijing Union University, Beijing 100101, China

ABSTRACT

Considering the situation of some practical factors such as energy saving of the nodes and the high density of distributing nodes in wireless sensor networks, a wireless sensor network (WSN) node optimal coverage method based on improved genetic algorithm and binary ant colony algorithm is proposed in this paper. The genetic algorithm and ant colony algorithm are improved and fused aiming at their disadvantages. The binary code expects a low intelligence of each ant, and each path corresponds to a comparatively small storage space, thus considerably improving the efficiency of computation. The optimal working node set is computed according to the max-coverage area of working sensor and the min-number of working sensor constraint conditions to optimize algorithm. The simulation results demonstrate that the proposed algorithm can converge at the optimal solution fast and satisfy the requirement of low node utilization rate and a high coverage rate, thus prolonging the network lifetime efficiently.

KEYWORDS

Wireless sensor network Node optimal coverage, Genetic algorithm, Binary ant colony algorithm

1. INTRODUCTION

A wireless sensor network is made up of a large number of sensor nodes which possess perceptual ability, calculation of capacity, and communication ability with self-organizing style. Sensor nodes complete the data collection and transmission together. It has wide application such as battlefield surveillance, environmental protection, intelligent home furnishing, and so on [1, 2, 3, 4, 5, 6]. However, sensor nodes usually adopt limited-capacity battery to provide energy due to the limit of volume and cost. The energy cannot regenerate or supply. So the coverage and energy consumption are the key problems of wireless sensor network research.

At present, many scholars have done a lot of research for the node location and coverage optimization algorithm in wireless sensor network, and some feasible and effective algorithms were proposed such as Huang algorithm [7], TianD algorithm [8], CCP algorithm [9], and so on [10, 11,12]. Chi-Fu Huang and Yu-Chee Tseng proposed a polynomial time algorithm (polynomial-time algorithms) in the literature [4]; in the sensor network, it is a distributed protocol method, which has positioning capabilities, environmental monitoring capability, and fault tolerant capability. Liang Ying, Zeng Peng, and Yu Haibin, in reference [7], proposed using adaptive cluster head communication (cluster-heads communication) method. The method can ensure that energy consumption is balanced in the whole network and can improve the network life cycle. Mihaela Cardei and Ding-Zhu Du, in the literature [8], proposed a rule and method to establish a "maximum disjoint set cover problem." The method is that nodes which are included in the maximum disjoint coverage area are at work, while other nodes are in a dormant state, which saves the system energy and prolongs the network life cycle. Jia Jie, Chen Jian, Chang Guiran, Zhao Linliang, and Wang Guangxing, in the literature [11], put forward the two based on weighted genetic algorithm and constraint genetic algorithm-based optimization coverage mechanism. According to generated fitness function to do genetic algorithm operation and calculate the sensor network, full coverage region is required for the approximation of the optimal set of nodes and complete working node selection, thereby reducing the redundant network and prolonging the survival time of the network. Xu et al. [13, 14, 15, 16, 17] proposed a crowdsourcing-based framework for processing mobile network data. Joon-Min Gil and Youn-Hee Han, in the literature [10], proposed a goal based on genetic algorithm coverage scheduling scheme, using evolutionary global search techniques for monitoring all targets and that can find the optimal coverage set, prolonging the network life cycle. Mohamed Aadhil M. and Roselin M.E.J., in the literature [12], using the genetic algorithm to optimize the node placement method, the method can achieve the purpose of optimizing the node coverage.

However, these algorithms have some limitations. The complexity of Huang algorithm is too high, and it could easily lead to the blockage of the channel when the communication data is large. The computing complexity and communication complexity of the TianD algorithm are lower, but there are two problems: first, inability to make the redundant nodes of monitoring area boundary dormant and second, without consideration of the overlap between node coverage areas, which led to elected excessive working nodes, thus affecting the lifetime of the network. CCP algorithm is not suitable for high-density sensor networks.

Based on the above issues and the special predominance of genetic algorithm and ant colony algorithm possess, a WSN node optimal coverage method based on improved genetic algorithm and binary ant colony algorithm is proposed in this paper, based on the location information of the nodes to build node coverage optimization model and then using the improved genetic algorithm and binary ant colony algorithm to solve the optimization model. The optimal solution is used as the working nodes of wireless sensor networks, the remaining nodes sleep to save energy. The simulation results show that this algorithm proposed in this paper effectively improves the coverage rate of the wireless sensor network.

2. WIRELESS SENSOR NETWORK COVERAGE OPTIMIZATION

2.1 Wireless sensor network coverage

The network coverage is the basic problem of the construction of wireless sensor networks. The network coverage needs to reach the maximum under the conditions that guarantee a certain quality of service (QoS) [18, 19, 20, 21], by measuring the network coverage to determine whether the presence of communication blind area and get the monitoring area of wireless sensor network coverage. The sensor node distribution can be adjusted or sensor nodes can be added according to the measurement condition. In addition, by changing the density of the network coverage, the more sensor nodes were deployed in some important areas in order to guarantee the reliability of measured data. Therefore, the network coverage is not only for communication coverage and monitoring area coverage but also for better meeting of the needs of certain applications. Generally speaking, quality of service is the measure of wireless sensor network coverage [22].

The wireless sensor network coverage is mainly divided into certainty coverage and the random covering. For the certainty coverage, the coverage area size is determined in order to cover the area using the minimum number of sensors. The certainty coverage is adopted to deploy the sensor nodes of a known region in this paper.

At present, a kind of widely used strategy is choosing part nodes which can be enough to cover the monitoring area as working nodes, at the same time, turn off redundant nodes [23]. A certain number of sensor nodes are deployed in the monitoring area, under the ensuring network connectivity in normal condition, letting part of the redundant nodes into low-power sleep state, while the rest of the active node is used to cover the whole region in the most likely small amount, which form the optimal coverage node set. In the wireless sensor network, the node number in the increased network not only means increased cost but also brings the limited bandwidth congestion because of the inter-node communication conflict. The optimal coverage node set is the minimum node set which does not affect the entire network coverage. This way does not only reduce potential conflicts in a wireless channel and reduce the possibility of competing for access to the media but also save energy in order to prolong the network lifetime [11].

2.2 Wireless sensor network optimization coverage model

Supposed that the target monitoring area A is a two-dimensional plane, and it will be divided into $m \times n$ grid, and each grid area is 1. The N sensor nodes are randomly distributed in the A, the sensor node set S

can be expressed as $S=(s_1, s_2, \cdots, s_i, \cdots, s_N)$. The (x_i, y_i) is the coordinate of sensor node s_i in a target monitoring area A, each node coordinate is known, and effective perception radii are $r$and $s_i = \{x_i, y_i, r\}$. Perception range of node s_i is a circular region with the coordinates (x_i, y_i) of node s_i as the center and the r as the perception radius, to express with $c_i = \{p \in A | d(p, (x_i, y_i)) \leq r, i \in [1, N]\}$, where, p is an arbitrary point of area A, and d is the Euclidean distance. The perception range of the whole sensor network is $C = \cup_{i \in [1,N]} c_i$, that all network nodes perceptual range are in union.

There is $m \times n$ target point. $(a_1, a_2, \cdots, a_j, \cdots a_{m \times n})$ need to be monitored or sensed in target monitoring area A; suppose a_j is s_i perception then $d(a_j, s_i) \leq r$; an arbitrary point of monitoring area A is at least perceived by a sensor node. Assume the communication range of sensor node s_i is a circle with the coordinates (x_i, y_i) of node s_i as the center and the r_c as the radius; if $d(s_i, s_j) \leq r_c$, then the sensor node s_i and s_j can communicate directly.

Definition 1: given a sensor node set S, the communication network $G=(V, E)$, $\{V=S, L(s_i, s_j) \in E\}$ composed of S is an undirected graph, if there is a communication path between any two nodes in the communication diagram G derived from S then the communication diagram is connected, when $d(s_i, s_j) \leq r_c$.

Definition 2: for the sensor node set S and the target monitoring area A; if every point of A is perceived by a sensor node of S then the S is a covering set of A. If the communication diagram derived from S is connected then the S is a connected cover set of A.

Definition 3: for the connected cover set S and the target monitoring area A, the minimal connected cover set is looking for a minimal subset $S' \subseteq S$, while the monitoring area A is covered by S' completely and the communication diagram derived from S' is connected.

On the basis of energy efficiency, this paper presents optimization goal of a connectivity covering set: ensuring that each target point in A is at least covered by a sensor node, at the same time, ensuring that a subset is a connected set.

In order to ensure connectivity, the communication radius r_c is at least two times perception radius r, namely $r_c \geq 2r$. On this condition, the researchers only need to consider the coverage problem in the sensor network, if the network is covered then it is connected [24].

Let $P_{cov}(x,y,s_i)$ represent the probability that any point (x,y) of monitoring area A is covered by sensor node $s_i(x_i,y_i)$ then

$$P_{cov}(x,y,s_i) = \begin{cases} 1, & (x-x_i)^2 + (y-y_i)^2 \le r^2 \\ 0, & \text{else} \end{cases} \tag{1}$$

As long as there is a sensor node, s_i covers the monitoring point (x,y) then the monitoring point (x,y) is covered by the sensor node set S. So the regional coverage area of sensor node set k is expressed as follows:

$$A_{area}(S) = \sum_{x=1}^{m}\sum_{y=1}^{n} P_{cov}(x,y,S)\,\Delta x\,\Delta y \tag{2}$$

The area of monitoring area is A_s, the working sensor node set is S', the target function of network coverage of working node set is as follows:

$$f_1(S') = A_{area}(S')/A_s \tag{3}$$

where $A_{area}(S')$ is the cover area of working sensor node set S'.

The target function of nodes utilization rate is as follows:

$$f_2(S') = |S'|/N \tag{4}$$

where N is the total number of sensor nodes, $|S'|$ is the number of working sensor nodes.

The aims of wireless sensor network node optimal coverage are to reduce the sensor node's utilization and to ensure maximum network coverage at the same time, this is a multi-objective optimization problem. So the multi-objective optimization coverage model of wireless sensor network is expressed as follows:

$$\max f(S') = \max\left(f_1(S'),\ 1-f_2(S')\right) \tag{5}$$

3. IMPROVED GENETIC ALGORITHM AND BINARY ANT COLONY ALGORITHM

3.1 Genetic algorithm

The genetic algorithm (GA) is a kind of self-adapting heuristic global search algorithm which was derived from imitating the thought of natural biological evolution. In nature, it is a cycle process made up of reproduction-crossover-mutation operators. In the process of searching for the global optimum solution, GA needs neither the information of gradient nor the calculus computing, it can find out the global optimum solution or near-optimal solution in the solution space with high probability only by operating the reproduction-crossover-mutation operators; thereby, it could reduce the probability of getting into the local minimum efficiently.

Adaptive genetic algorithm (AGA) is a kind of GA that has scale reproduction and self-adaptive crossover and mutation operations. In the process of searching for the optimum parameter, AGA changes the crossover probability and mutation probability adaptively according to the different conditions of individuals in order to keep the diversity of colony and prevent the premature convergence; further, it can enhance the calculating speed and precision of the algorithm.

$$P_c = \begin{cases} k_1(f_{\max}-f')/(f_{\max}-f_{avg}) & if \;\; f' > f_{avg} \\ k_3 & if \;\; f' < f_{avg} \end{cases} \tag{6}$$

$$P_m = \begin{cases} k_2(f_{\max}-f)/(f_{\max}-f_{avg}) & if \;\; f > f_{avg} \\ k_4 & if \;\; f < f_{avg} \end{cases} \tag{7}$$

where P_c is exchanging probability, P_m is mutation probability, $f_{\max}$ is the biggest fitness of colony, f_{avg} is the average fitness of colony, f' is the bigger fitness of two strings used for crossover, f is the fitness of the individual to mutate.

3.2 Binary ant colony algorithm

The ant colony algorithm is a kind of new simulated evolutionary algorithm. Ants will release a peculiar pheromone in its path when they are searching for food, which make other ants feel the kind of peculiarity and tend towards the direction of high density [25]. Ant colony algorithm converge the optimal path through the accumulation and update of pheromone and achieve optimization through the appropriate times of iteration.

The binary ant colony algorithm denotes the parameters directly with binary-coding; thus, it can take up less storage space and reduce the complexity of the algorithm compared to decimal-coding. Moreover, the traversal search capability of binary-coding is better than that of decimal-coding.

Definition of the directed graph $G = (V, R)$, the vertex set

$$\left\{ \{v_0(c_s),\ v_1(c_N^0),\ v_2(c_N^1),\ v_3(c_{N-1}^0),\ v_4(c_{N-1}^1),\ \cdots, \right.$$

V is:

$$\left. v_{2N-3}(c_2^0),\ v_{2N-2}(c_2^1),\ v_{2N-1}(c_1^0),\ v_{2N}(c_1^1) \right\} \},$$

where c_s is the start vertex, c_j^0, c_j^1, respectively, stand for the 0 and 1 states of bit b_j in binary-coding. For all vertex, when $j = 2, 3, \cdots, N$, there are only two directed arcs which point to c_{j-1}^0 and c_{j-1}^1, these two directed arcs respectively stand for the 0 and 1 states of each ant next step allowed, the N is the binary encoding length, this binary ant colony network is shown in Fig. 1. Each ant travels their own path and then get the solution of the problem by integrating each ant's traversal solution [26, 27, 28, 29, 30, 31].

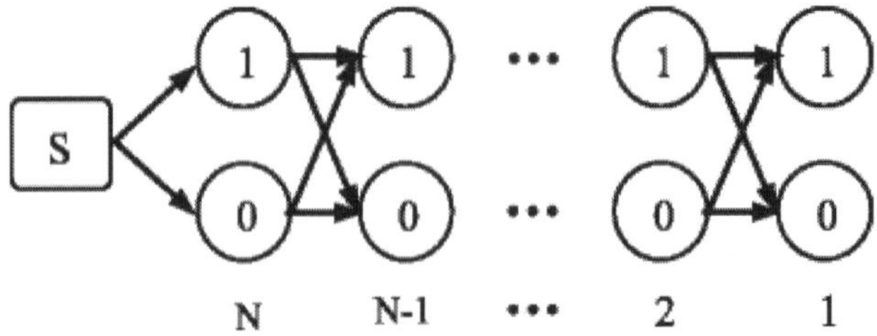

Figure 1. Binary ant colony network

At the start, the information quantity of each path is equal, suppose $\tau_{i,j}(0) = C$ (C is a constant), $\Delta\tau_{i,j} = 0$ $(i,j = 1, 2, \cdots, N)$. In the process of movement, the k ($k = 1, 2, \cdots, M$) ant determines the movement direction according to the information quantity of each path, and the movable probability is

$$p_{i,j}^{k} = (\tau_{i,j}^{\alpha}(t) \cdot \eta_{i,j}^{\beta}(t)) / (\sum_{s \in \text{allowed}_k} \tau_{i,s}^{\alpha}(t) \cdot \eta_{i,s}^{\beta}(t)) \tag{8}$$

where m is the number of ants, $p_{i,j}^{k}$ is the probability that the k ant moves from the i position to the $j$position at the t moment. α ($\alpha \geq 0$) is the relative importance of a moving path. β ($\beta \geq 0$) is the relative importance of visibility. $\tau_{i,j}$ is the residual information amount in the ij connection at the t moment. $\eta_{i,j}$is the visibility in the ij connection. $\text{allowed}_k = \{0, 1\}$ is the next select status.

With the passage of time, the leaving information gradually disappear. ρ $(0 \leq \rho \leq 1)$ is persistence of a moving path. $1 - \rho$ is the information loss degree. After n moment, the ant completes one cycle. The $\tau_{i,j}$ is renewed as follows:

$$\tau_{i,j}(t+1) = \rho \cdot \tau_{i,j}(t) + \Delta\tau_{i,j} \tag{9}$$

where $\Delta\tau_{i,j} = 1/f_{\text{best}}(S)$, $f_{\text{best}}(S)$ is the iteration-best solution or global best solution, which can enhance the optimizing ability and improve the speed of solution.

Compared with the traditional ant colony algorithm, the biggest difference of binary ant colony algorithm is the ant's path selection. The states of binary ant colony algorithm only have 0 and 1, the ant does not need to remember those nodes which have traveled, it selects only on the basis of prior to the two path pheromones. The traverse path in the binary ant colony algorithm is shown in Fig. 2.

3.3 Improved genetic algorithm and binary ant colony algorithm

The improved genetic algorithm and binary ant colony algorithm introduce a genetic operator into the binary ant colony algorithm, using the crossover and mutation operations to expand the search space and enhance the global property of solutions [26, 32]. The solution is formed by the binary ant colony algorithm as an initial population of genetic algorithm then a better solution is searched by iterative evolution. Thus, it can enhance the convergence speed of algorithm, improve efficiency for solution, and avoid the local optimal and precocious defects [32].

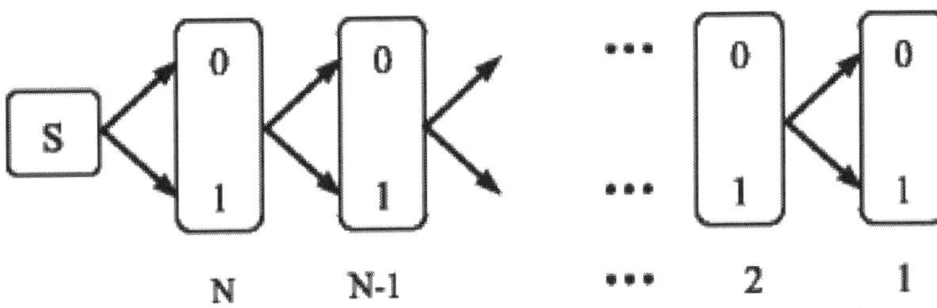

Figure 2. Traverse path in the binary ant colony algorithm

Randomly generated initial solution executes multiple evolutionary using the reproduction-crossover-mutation operators of genetic algorithm and, at the same time, using the solution obtained by genetic algorithm to initialize the information pheromones of a network. If the loop end conditions are met, then the algorithm ends; else, the researchers execute the ants' traverse and pheromone update and reserve the optimal solution, execute the reproduction-crossover-mutation operation to the obtained solution set and reserve the optimal solution, and use the optimal update pheromone and determine whether the end condition is met.

The algorithm flow is shown in Fig. 3.

4. WIRELESS SENSOR NETWORKS NODES OPTIMAL COVERAGE BASED ON IGA-BACA

The wireless sensor network node optimal coverage is a typical multi-objective optimization problem, this problem can be expressed as follows: existing sensor set $S = (s_1, s_2, \cdots, s_i, \cdots, s_N)$, seeking the working set S', which can obtain the maximum network coverage rate, and the minimum working node set. When the maximum network coverage rate is selected as the optimal objective, network sensor nodes with hexagon structure distributed in the monitoring area, the minimum working nodes set can be obtained [33].

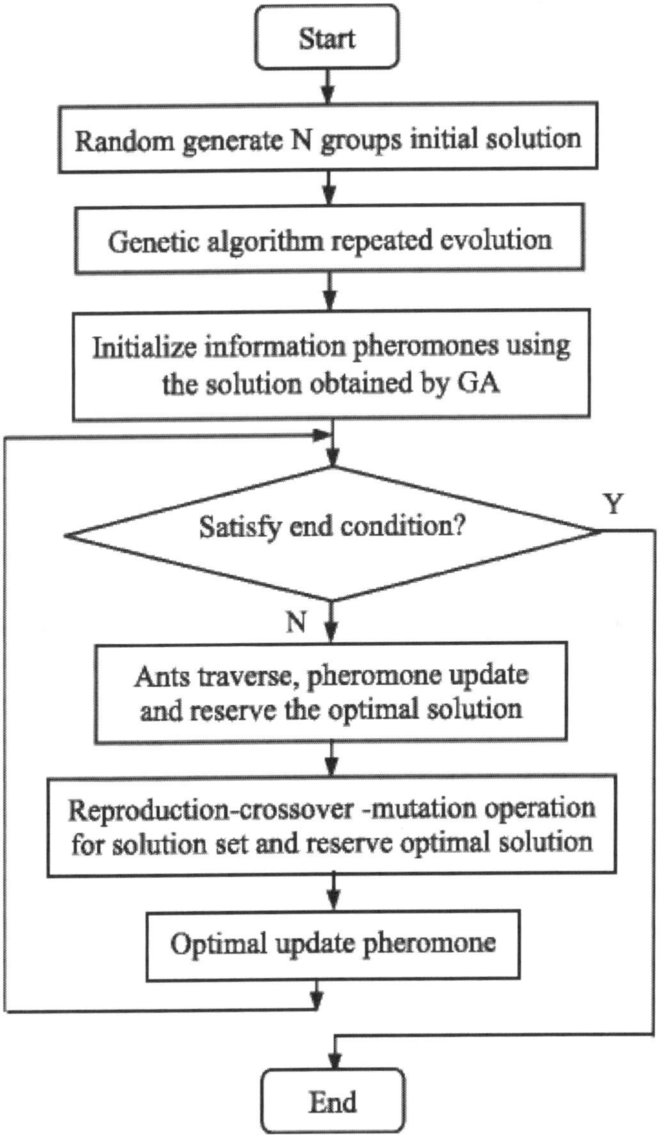

Figure 3. Flow chart of improved genetic algorithm and binary ant colony algorithm

From formula (5), we know that the maximum network coverage $f_1(S')$ and minimum working nodes set $f_2(S')$ are contradictory; compromising the two, the multi-objective optimization coverage model of wireless sensor network is changed into the maximum objective function $f(S')$.

$$f(S') = \left(f_1(S') \cdot f_1(S')\right)/f_2(S') \tag{10}$$

The algorithm implementation of wireless sensor network node optimal coverage based on improved genetic algorithm and binary ant colony algorithm is as follows [34, 35, 36, 37, 38, 39].

4.1 Code

It is a key problem to complete the mapping from the solution space of optimization problem to the coding space. Aiming at the random distribution nodes, the control vector (individual) $L = (l_1, l_2, \cdots, l_i, \cdots, l_N)$ of wireless sensor node is expressed in a binary coded form, which stands for the position of all sensor nodes in the target monitoring area *A*, where the l_i value is 1 or 0, which separately correspond to the active or sleep of the *i* sensor node. The selection of nodes and the gene of chromosome is one to one correspondence.

4.2 Initial solution and adaptation function

At random initialization, ten sensor nodes randomly distribute in the target monitoring area *A*, where the nodes of 1, 2, 3, 5, 7, 8, 10, 11, 12, 15 are working nodes and then the corresponding individual is { 1 1 1 0 1 0 1 1 0 1 1 1 0 0 1 }, this is shown in Fig. 4. A large number of individuals will be generated in the initialization phase, which is called a colony.

Use the formula (10) as the adaptation function, the higher is the adaptation function value, the higher is the selected probability of individual.

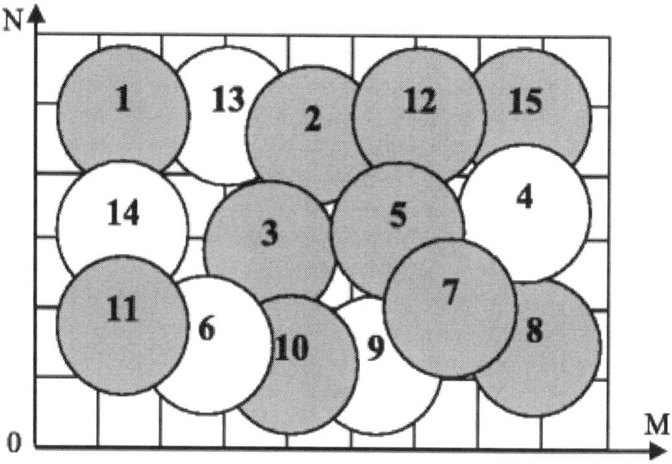

Figure 4. Initialize individual { 1 1 1 0 1 0 1 1 0 1 1 1 0 0 1 }

4.3 Operation operator

The operation operator of genetic algorithm and binary ant colony algorithm is composed of four operators that is reproduction operator, crossover operator, mutation operator, and pheromone update operator.

The reproduction operator reproduces the individuals to the new colony according to the probability in proportion as their adaptive value. After reproduction, the preponderant individuals are preserved and the inferior individuals are weeded out, and the average fitness degree of the colony is increased, but the variety of colony is lost at the same time. The action of reproducing an operator is to realize the principle of winner priority for preserving predominance and natural selection and making the colony converge on the optimum solution.

The crossover operator first selects two individuals stochastically according to the certain adaptive exchanging probability P_c then it can produce two new individuals by exchanging parts of chromo gene stochastically. The two-point crossover is adopted in this paper, randomly generating two crossover points is shown in Fig. 5. The genetic algorithm can generate filial generation colony which have higher average fitness and better individuals through the reproduction and crossover operators and can make the evolutionary process proceed to the optimum solution.

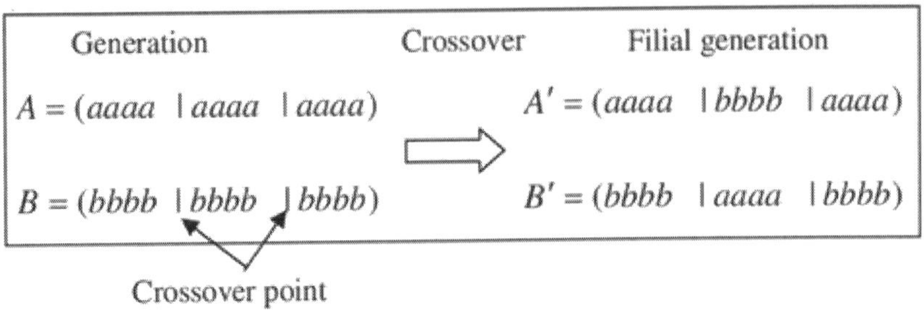

Figure 5. Two-point crossover

The mutation operator changes several bits of the chromosome string stochastically with a small probability P_m, namely turning 0 to 1 and 1 to 0. The mutation operator is very important to recoup the loss of colony diversity.

The pheromone update operator (T_g) is a mapping of pheromone update for optimal individual which is selected from an ant sequence. According to the Max-Min rule, pheromone only be released in the path of optimal solution ant traverse, namely $\Delta\tau_{i,j} = 1/f_{\text{best}}(S)$, $f_{\text{best}}(S)$ is the iteration-best solution or global best solution. The probability of pheromone update operator is shown as follows:

$$P\{T_g = x_i\} = f(x_i)/\sum_{k=1}^{N} f(x_k) \tag{11}$$

where, N is the number of individuals.

4.4 Ants traverse

When an ant travels, the path selection of different ants is mutual independence. Starting from the nest, every ant performs path selection with the movable probability formula (8) and according to Fig. 2; because of the binary-coding, the ants do not possess memory function, it selects only on the basis of prior to the two path pheromones. The path that has higher density pheromone will have higher selection probability. When an ant completes the path selection in an ant sequence, it arrives at the food source.

5. SIMULATION EXPERIMENT

The set of the monitoring area A is 100 m × 100 m, the perception radius r of sensor nodes is 10 m. When the network sensor nodes with hexagon structure distributed in the monitoring area and the distance d between nodes are $d = \sqrt{3} \cdot r$, the coverage of target monitoring area A can achieve optimal. From this, we know that 45 nodes with a radius of 10 can completely cover the monitoring area A, this is shown in Fig. 6.

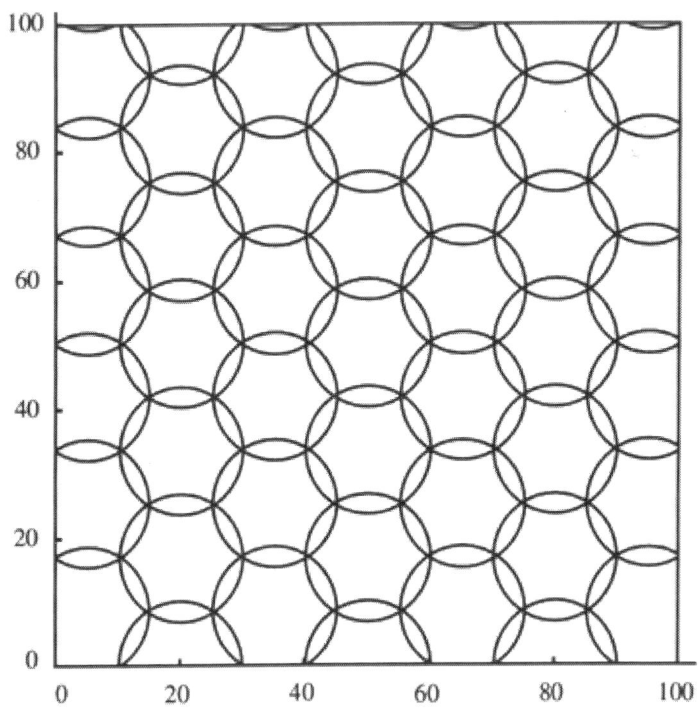

Figure 6. Node optimal coverage

We write the corresponding algorithm program using C++. The improved genetic algorithm and binary ant colony algorithm adopt two-point crossover, scale reproduction, and self-adaptive crossover and mutation operations. Population size is 60. There was random distribution of 100 sensor nodes in monitoring area A, the perception radius of sensor nodes is 10 m then every individual is composed of 100 binary bits. Algorithm iterative number is 200. Figure 7 is the 100 nodes in the area *A* random distribution. Although the nodes can cover the target region, but there are a lot of redundant nodes. Figure 8a–d shows the working nodes set coverage using the improved genetic algorithm and binary ant colony algorithm evolution 50, 100,150, 200 generations.

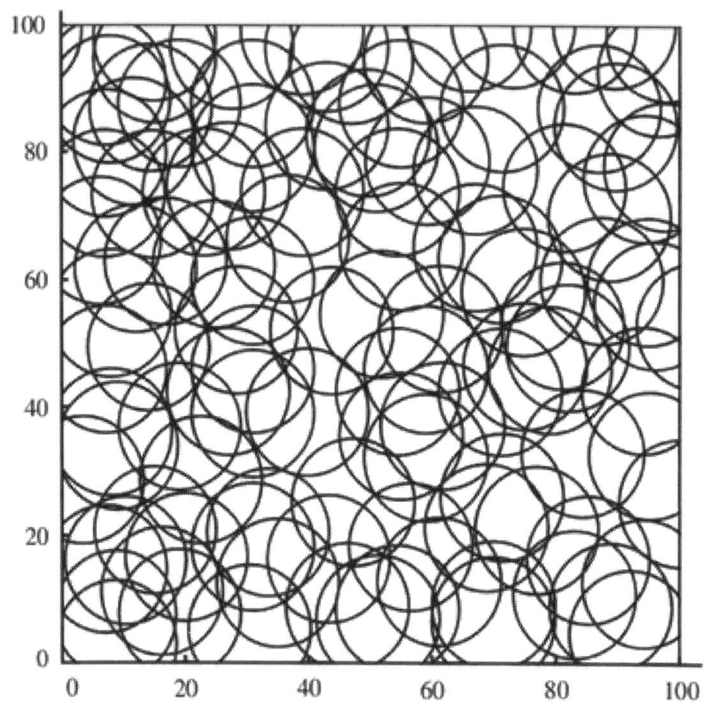

Figure 7. Random distribution 100 nodes in the area *A*

From Fig. 8a–d, we can see that the algorithm search towards the global optimal solution direction. When the iterative number is 200, the sensor nodes is near to the optimal distribution, using less number of sensor nodes, obtaining more than 97% coverage.

In order to evaluate the performance of IGA-BACA algorithm, in the other parameters unchanged, we evaluate its performance using different random deployment scenarios. In different sensor node density experiments, the number of sensor nodes deployed is, respectively, 100, 120, 150, 180, and 200. From the network coverage rate and the convergence of the algorithm, the improved genetic algorithm and binary ant colony algorithm are compared with ACA algorithm and GA algorithm. The experimental results are the average of 30 experiments. Figure 9 is a network coverage rate comparison chart of different sensor nodes.

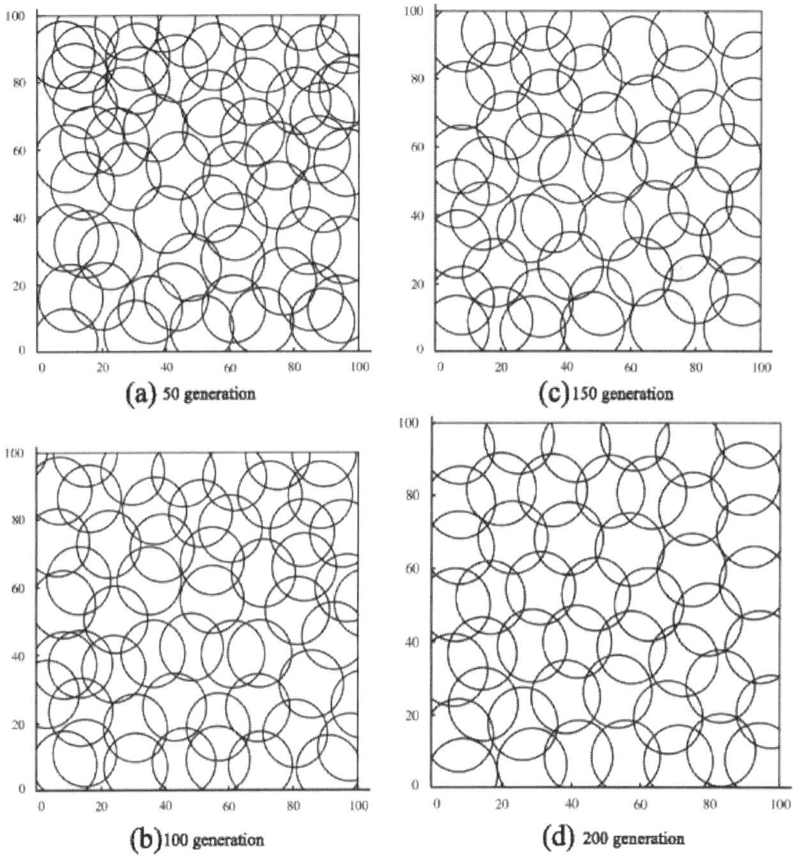

Figure 8. Working set node coverage by IGA-BACA evolution. **a** 50 generations, **b** 100 generations, **c** 150 generations, **d** 200 generations

From Fig. 9, we can see that the coverage quality increases along with the increase of sensor node density. In the same iteration, coverage quality of IGA-BACA algorithm is higher than those of GA algorithm and ACA algorithm. After 200 generations of evolution, the number of network sensor node is 100, the network coverage rate of IGA-BACA algorithm is 97.5 %, the network coverage rate of GA algorithm is 96.2 %, and the network coverage rate of ACA algorithm is 96.5 %. The number of network sensor node is 150, the network coverage rate of IGA-BACA algorithm is 98.7 %, the network coverage rate of GA algorithm is 97.3 %, and the network coverage rate of ACA algorithm is 97.7 %. The number of network sensor node is 200, the network coverage rate of IGA-BACA algorithm is 99.5 %, the network coverage rate of GA algorithm is 98.1 %, and the network coverage rate of ACA algorithm is 98.6 %. These results show that the IGA-BACA algorithm is better than the ACA algorithm and GA algorithm in wireless sensor network coverage quality improvement.

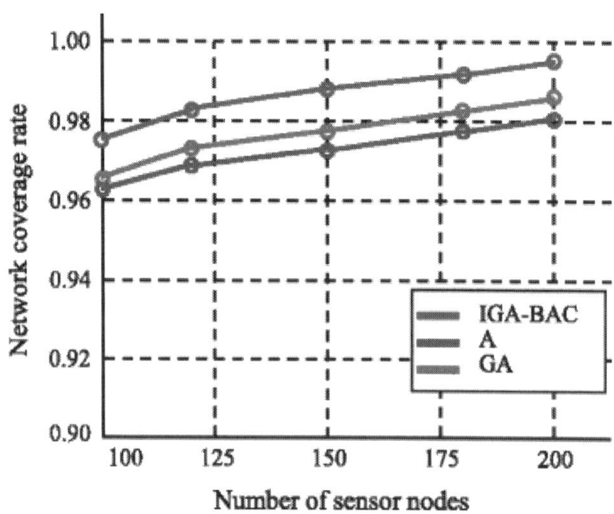

Figure 9. Network coverage rate comparison chart

Figure 10 is the convergence comparison chart of IGA-BACA algorithm, GA algorithm, and ACA algorithm in 200 generations of the evolutionary process. Figure 10a is the convergence comparison chart in the 100 sensor nodes deployed, and the Fig. 10b is the convergence comparison chart in the 200 sensor nodes deployed. Figure 10 shows the convergence of the three algorithms. From Fig. 10a, we can see that this figure clearly shows the convergence of the three algorithms' performance. The IGA-BACA algorithm, GA algorithm, and ACA algorithm can converge, and the convergence of IGA-BACA algorithm is better than that of ACA algorithm and GA algorithm, the convergence of

ACA algorithm is better than that of GA algorithm. From Fig. 10b, we can get the same conclusion, for the three algorithms, and the convergence of IGA-BACA is the best, followed by ACA, finally GA. So the performance of IGA-BACA is better than the GA algorithm and ACA algorithm under the same parameters.

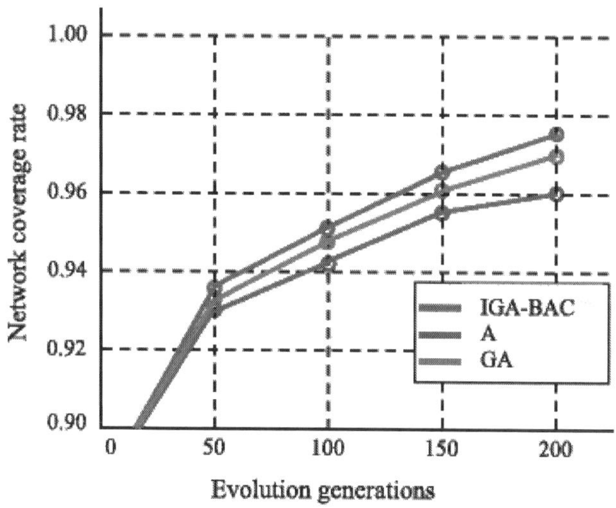

(a) The convergence of 100 sensor nodes

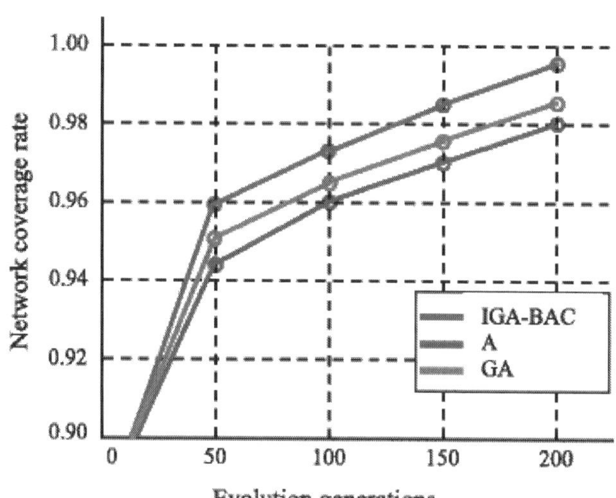

(b) The convergence of 200 sensor nodes

Figure 10. The algorithm convergence comparison chart. **a** The convergence of 100 sensor nodes. **b** The convergence of 200 sensor nodes

In order to further illustrate the validity of the IGA-BACA, we adopt IGA-BACA and GA to realize the optimal coverage of wireless sensor network node in the same conditions. In the 100 sensor nodes deployed, the simulation results by IGA-BACA and GA are shown in Table 1.

From Table 1, we can see that the network coverage rate of IGA-BACA is greater than that of GA in the same evolution generation condition. Moreover, the working set node number of IGA-BACA is smaller than that of GA in the same evolution generation condition. We also can see that along with the evolution generation increase, the network coverage rate is gradually approaching the optimal solution; moreover, the working set sensor node number is gradually reduced, this will reduce energy loss of a wireless sensor network and prolong the network lifetime. From the CPU time, we can see the IGA-BACA algorithm to optimize the speed faster than the GA algorithm.

All of these show that the IGA-BACA is superior to the GA and ACA at the convergence of the optimal solution with less working nodes and high network coverage rate.

6. CONCLUSIONS

The coverage and energy consumption are the key problems of wireless sensor network research because the sensor node energy cannot regenerate or supply. This paper analyzes the characteristics of wireless sensor network limited node energy. With the network coverage and the number of working set nodes as the optimization object, research on the working node set selection optimization problem of nodes explore the high-density deployment in wireless sensor network.

The wireless sensor network node optimal coverage based on improved genetic algorithm and binary ant colony algorithm is proposed and applied to node optimal coverage. The aim at the disadvantages of the genetic algorithm and ant colony algorithm is introduce binary ant colony algorithm and combine it with genetic algorithm to form improved genetic algorithm and binary ant colony algorithm which has higher accuracy and faster convergence speed. Considering the relationship between network coverage and the number of working set nodes, the complex multi-objective problem is simplified into a single-objective fitness function in this paper, which can balance network coverage rate and energy consumption of the system.

Table 1. Simulation results by IGA-BACA and GA

Evolution generations	IGA-BACA			GA		
	Network coverage rate (%)	Working set node number	CPU-time (s)	Network coverage rate (%)	Working set node number	CPU-time (s)
-	100	45	-	100	45	-
50	93	63	11	93.1	65	13
100	95.2	56	21	94.3	59	25
150	96.5	51	31	95.6	53	37
200	97.5	48	42	96.2	50	49

The simulation results show that the improved genetic algorithm and binary ant colony algorithm can achieve the optimal working set with less working nodes and high network coverage rate, effectively reducing energy loss of a wireless sensor network and prolonging the network lifetime.

ACKNOWLEDGEMENTS

This work is supported in part by the National Natural Science Foundation of China under Grant No. 61271370, the National High Technology Research and Development Program of China (863 Program) under Grant No. 2013AA013202, and Funding Project for Academic Human Resources Development in Beijing Union University No. 11101501105.

REFERENCES

1. IF Akyildiz, W Su, Y Sankarasubramaniam, E Cayirci, Wireless sensor networks: a survey. Comput. Netw.**38**, 393–422 (2002)

2. I.T. Almalkawi, Guerrero Zapata, M., Al-Karaki, J.N. A cross-layer-based clustered multipath routing with QoS-aware scheduling for wireless multimedia sensor networks. Int. J. Distributed Sensor Net. **2012**, 1–11 (2012)

3. J Yang, C Zhang, X Li, Y Huang, S Fu, MF Acevedo, Integration of wireless sensor networks in environmental monitoring cyber infrastructure. Wirel. Netw **16**(4), 1091–1108 (2010)

4. H Chi-Fu, T Yu-Chee, The coverage problem in a wireless sensor network. Mobile Network Appl. **10**(4), 519–528 (2005)

5. J Yick, B Mukherjee, D Ghosal, Wireless sensor network survey. Comput. Netw. **52**, 2292–2330 (2008)

6. H Alemdar, C Ersoy, Wireless sensor networks for healthcare: a survey. Comput. Netw. **54**, 2688–2710 (2010)

7. Y LIANG, P Zeng, H Yu, Energy adaptive cluster-head selection for wireless sensor networks. Inf. Control.**35**, 141–146 (2006)

8. M Cardei, D Ding-Zhu, Improving wireless sensor network lifetime through power aware organization. Wirel. Netw **11**, 333–340 (2005)

9. X. R. Wang, G. L. Xing, Y. F. Zhang, C. Y. Lu, R. Pless, C. D. Gill. Integrated coverage and connectivity configuration in wireless sensor networks. Proceedings of the 1st ACM Conference on Embedded Networked Sensor Systems (SenSys'03), Los Angeles, CA, 2003.

10. G Joon-Min, H Youn-Hee, A target coverage scheduling scheme based on genetic algorithms in directional sensor networks. Sensors, Sensor **11**, 1888–1906 (2011).

11. J Jia, J Chen, J Chang, L Zhao, G Wang, Optimal coverage scheme based on genetic algorithm in wireless sensor networks. Control Decis. **1**(22), 1289–1292 (2007)

12. MM Islam, M Ahasanuzzaman, MA Razzaque, MM Hassan, A Alelaiwi, Y Xiang, Target coverage through distributed clustering in directional sensor networks. EURASIP J. Wireless Commun. Netw. **3**, 1–18 (2015)

13. X Luo, Z Xu, J Yu, X Chen, Building association link network for semantic link on web resources. IEEE Trans. Autom. Sci. Eng. **8**(3), 482–494 (2011)

14. Z. Xu et al. Crowdsourcing based social media data analysis of urban emergency events. Multimed Tools Appl.

15. Z. Xu et al. Crowdsourcing based description of urban emergency events using social media big data. IEEE Transactions on Cloud Computing.

16. Z Xu et al., Participatory sensing based semantic and spatial analysis of urban emergency events using mobile social media. EURASIP J. Wirel. Commun. Netw. **44**, 2016 (2016)

17. Z. Xu et al. Building knowledge base of urban emergency events based on crowdsourcing of social media. Concurr. Comput.

18. HM Ammari, On the problem of k-coverage in mission-oriented mobile wireless sensor networks. Comput. Netw. **56**(7), 1935–1950 (2012)

19. E. Lattanzi, A. Bogliolo, VirtualSense. A java-based open platform for ultra-low-power wireless sensor nodes. Int. J. Distributed Sensor Netw. **2012**, 1–16 (2012)

20. C Wenping, C Si, L Deying, Minimum-delay POIs coverage in mobile wireless sensor networks. EURASIP J. Wireless Commun. Netw **2013**, 1–12 (2013)

21. Y Sun, Y Sun, Z Fang, Connectivity and coverage of wireless sensor networks. J. Tianjin University **38**(1), 14–17 (2005)

22. S. Meguerdichian, F. Koushanfar, M. Potkonjak et al, Coverage problem in wireless ad-hoc sensor networks. IEEE INFOCOM, vol 3 (IEEE Press, IEEE, Anchorage, 2001) 1380-1387.

23. M Liu, J Cao, Y Zheng, L Chen, L Xie, Analysis for multi-coverage problem in wireless sensor networks. J. Softw. **18**, 127–136 (2007)

24. H. Do. Research on the coverage and connectivity algorithm of wireless sensor network, Master degree thesis, Hunan University, (2007) p. 41–45.

25. Y Sun, JW Tian, WSN path optimization based on fusion of improved ant colony algorithm and genetic algorithm. J. Comput. Info. Syst. **6**, 1591–1599 (2010)

26. W Xiong, P Wei, Binary ant colony evolutionary algorithm. Acta Automat. Sin. **33**, 259–264 (2007)

27. B Yan, W Xiong, M Cheng, Q Ye, Multi-population binary ant-colony algorithm with congestion control strategy. Control Theory & Applications **26**(4), 387–394 (2009)

28. Q Qian, M Cheng, W Xiong, M Zhou, Reviews of binary ant colony optimization. Appl. Res. Computers**29**(4), 1211–1215 (2012)

29. C Yan, W Xiong, Y Zhang, Binary ant colony optimization based on traffic flow control. J. Syst. Simulation**19**(10), 2346–2350 (2007)

30. W Xiong, P Wei, J Zhao, Binary ant colony algorithm with signal transfer. Pattern Recognit. Artif. Intell.**20**(1), 15–20 (2007)

31. G Hu, W Xiong, X Zhang, J Yuan, Binary ant colony algorithm with controllable search bias. Contr. Theor. Appl. **28**(8), 1071–1080 (2011)

32. J Pan, X Wang, Y Chen, Improved ant colony algorithm for mobile robot path planning. J. China University of Mining & Technology **41**(1), 108–113 (2012)

33. Z Honghai, JC Hou, Maintaining sensing coverage and connectivity in large sensor networks. Ad- hoc and Sensor Wirel. Netw **1**, 89–124 (2005)

34. Y Zeng, J Cao, J Hong, S Zhang, L Xie, Secure localization and location verification in wireless sensor networks: a survey. J. Supercomput. **64**(3), 685–701 (2013)

35. WSH Siu, C-K Chan, HCB Chan, Green cargo routing using genetic algorithms. Lect. Notes Eng. Comput. Sci. **1**, 170–175 (2012)

36. A Chen, S Kumar, TH Lai, Local barrier coverage in wireless sensor networks. IEEE Trans. Mob. Comput.**9**(4), 491–504 (2010)

37. X. Yu, J. Zhang, J. Fan, and T. Zhang. A faster convergence artificial bee colony algorithm in sensor deployment for wireless sensor networks. Int. J. Distributed Sensor Netw. **2013**, 1–9 (2013).

38. M Koç, I Korpeoglu, Coordinated movement of multiple mobile sinks in a wireless sensor network for improved lifetime. EURASIP J. Wireless Commun. Netw. **2015**, 1–16 (2015)

39. S Jin-Taek, L Heung-No, 4-ary network coding for two nodes in cooperative wireless networks: exact outage probability and coverage expansion. EURASIP J. Wireless Commun. Netw. **2012**, 1–12 (2012)

4

DYNAMIC DIFFERENTIAL MODULATION OF SUB-CARRIERS IN OFDM

Charles U. Ndujiuba, Augustus E. Ibhaze

Electrical & Information Engineering, Covenant University, Ota, Nigeria

ABSTRACT

Orthogonal Frequency Division Multiplexing (OFDM) is a special case of FDM, which is implemented in digital communication systems. In this technology, multiplexing is applied to independent signals, which are sub-sets of one main signal, and applied to independent channels (sub-carriers) of a bandwidth. Here a user utilizes all sub- carriers to transmit its data. The questions here are: In OFDM can we modulate each sub-carrier differently from one another to enhance the overall integrity of the transmitted signal? If so, what determines the type of modulation technique that is suitable for the individual channels (sub-carriers)? These are the questions we seek to answer in this work, using simulation results to illustrate our theoretical analysis. We use bit error rate (BER) and data throughput as the metric for measuring the performance of the system.

KEYWORDS

FDM, OFDM, Sub-carriers, BER, Channel State Information (CSI), QPSK, QAM

1. INTRODUCTION

These days communication requires a very high rate with high reliability. Two major difficulties that hinder reliable communication via high rate wireless communication systems are bandwidth limitation of communication channels and multipath fading. To surmount these difficulties in analogue transmission, the basic principle of multi-carrier transmission, can be used, accomplished by means of Frequency Division Multiplexing (FDM) technology.

In FDM, independent signals, produced by different users are carried by independent sub-carriers of a given bandwidth. Each user or signal source is allocated a channel (sub-carrier) for the duration of the transmission, and each channel can be modulated by a modulation technique different from the other channels, without affecting their quality of performance. The modulation technique is usually customized to the type of information.

In analog transmission, signals are commonly multiplexed using frequency-division multiplexing (FDM), in which the carrier bandwidth is divided into sub-channels of different frequency widths, each carrying an independent signal at the same time in parallel. Each individual channel occupies a finite frequency range, typically some multiple of a given base frequency. Traditional terrestrial microwave and satellite links employ FDM. Although FDM in telecommunications is being reduced, several systems will continue to use this technique, namely: broadcast & cable TV, and commercial & cellular radio.

Figure 1 illustrates the main concepts of a typical FDM scheme. This process can continue until the available bandwidth on the cable or microwave link is exhausted.

The RF spectrum must be shared, yet every day there are more users for that spectrum as demand for communications services increases, hence the increase in the quest for better spectral efficiency. This can be achieved by employing digital modulation schemes, because they have greater capacity to convey large amounts of information than analog modulation schemes.

OFDM is a digital multi-carrier modulation scheme in which a single high rate data stream is divided into multiple low rate data streams. These data streams are then modulated using subcarriers which are orthogonal to each other. In this way the symbol rate on each sub channel is greatly reduced, and hence the effect of inter symbol interference (ISI) due to channel dispersion in time caused by multipath delay spread is reduced. Each subcarrier is modulated with a conventional digital modulation scheme (such as QPSK, 16QAM, etc.) at low symbol rate.

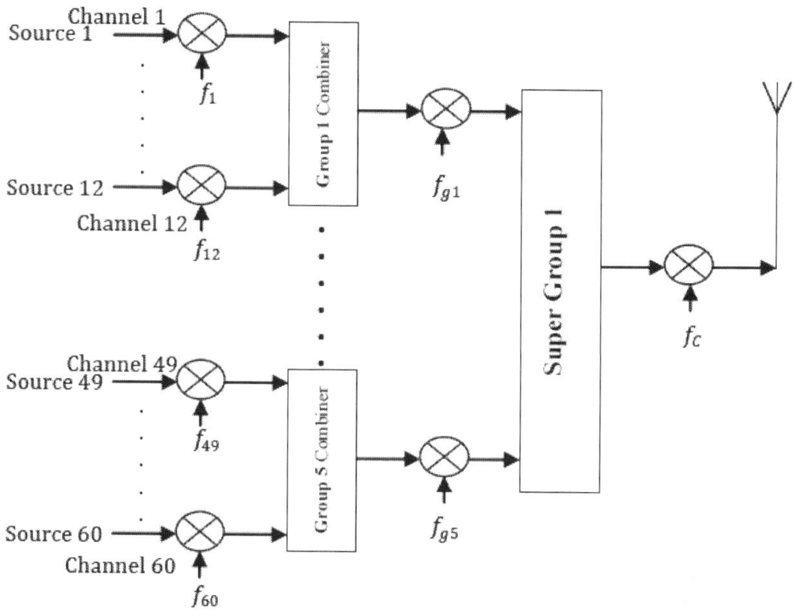

Figure 1. FDM Scheme

Table 1. Capacity of an FDM scheme

Analogue Carrier System			
No. of Voice Channels	Bandwidth	Spectrum	Terminology
12	48 KHz	60-108 KHz	Group
60	240 KHz	312-552 KHz	Supergroup
600	2.520 MHz	564-3084 KHz	Mastergroup

The low-pass equivalent OFDM signal is expressed as: [1-3]

$$\nu(t) = \sum_{k=0}^{N-1} X_k e^{j2\pi kt/T}, \quad 0 \leq t < T,$$

where $\{X_k\}$ are the data symbols, N is the number of sub-carriers, and T is the OFDM symbol time. The sub-carrier spacing of $\frac{1}{T}$ makes them orthogonal over each symbol period; this property is expressed as: [1-3]

$$\frac{1}{T}\int_0^T \left(e^{j2\pi k_1 t/T}\right)^* \left(e^{j2\pi k_2 t/T}\right) dt$$
$$= \frac{1}{T}\int_0^T e^{j2\pi (k_2 - k_1)t/T} dt = \delta_{k_1 k_2}$$

where $(\cdot)^*$ denotes the complex conjugate operator and δ is the Kronecker delta.

Since each subcarrier has a lower information rate, the data symbol periods in a digital system will be longer, adding some additional immunity to impulse noise and reflections.

The OFDM scheme differs from traditional FDM in the following interrelated ways:

1) Multiple carriers (called subcarriers) carry one information stream, and

2) The subcarriers are orthogonal to each other.

Figure 2 illustrates the concepts of OFDM and its similarity with the FDM scheme [4, 5]. The high data rate serial input bit stream is fed into serial to parallel converter to get low data rate output parallel bit stream. Input bit stream is taken as binary data. The low data rate parallel bit stream is modulated in Signal Mapper. Modulation can be BPSK, QPSK, etc. The OFDM symbols are fed to Inverse Fast Fourier transform so that each subcarrier is assigned with a specific frequency. In this block, orthogonally in subcarriers is introduced. In IFFT, the frequency domain OFDM symbols are converted into time domain OFDM symbols. This OFDM signal is allowed to pass through digital to analogue converter (DAC). In DAC the OFDM signal is fed to RF power amplifier for transmission. If there is frequency mismatch between transmitter and receiver local oscillators, frequency offset occurs. Then the signal is allowed to pass through additive white Gaussian noise channel (AWGN channel) [5].

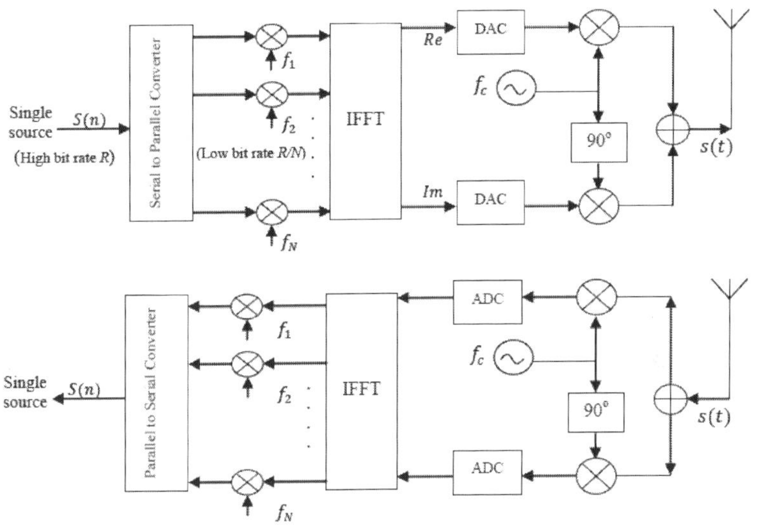

Figure 2. OFDM Scheme

An OFDM carrier signal is the sum of a number of orthogonal sub-carriers, with baseband data on each sub-carrier being independently modulated commonly using some type of quadrature amplitude modulation (QAM) or phase-shift keying (PSK). This composite baseband signal is typically used to modulate a main RF carrier. S[n] is a serial stream of binary digits. By inverse multiplexing, these are first demultiplexed into parallel streams, and each one mapped to a (possibly complex) symbol stream using some modulation constellation (QAM, PSK, etc.). Note that the constellations may be different, so some streams may carry a higher bit-rate than others [6, 7].

An inverse FFT is computed on each set of symbols, giving a set of complex time-domain samples. These samples are then quadrature-mixed to pass band in the standard way. The real and imaginary components are first converted to the analogue domain using digital-to-analogue converters (DACs); the analogue signals are then used to modulate cosine and sine waves at the carrier frequency, fc, respectively. These signals are then summed to give the transmission signal, s[t] [7].

In an OFDM transmission system, each subcarrier is attenuated individually under the frequency-selective and fast fading channel. The channel performance may be highly fluctuating across the subcarriers and varies from symbol to symbol [6].

If the same fixed transmission scheme is used for all OFDM subcarriers, the error probability is dominated by the OFDM subcarriers with highest attenuation resulting in a poor performance. Therefore, in case of frequency selective fading the error probability decreases very slowly with increasing average signal-to-noise ratio (SNR). This problem can be mitigated if each subcarrier can be allocated a different modulation scheme based on the measured channel conditions. This allows subcarriers to be dynamically allocated modulation schemes based on the SNR of each subcarrier. Those subcarriers with a low SNR can be allocated to use BPSK (1 b/s/Hz) or to transmit no data at all. Subcarriers with a high SNR can transmit higher modulation schemes such as 256-QAM (8 b/s/Hz) allowing a higher system throughput [8].

This will substantially improve the performance and data throughput of an OFDM system. For example if the subcarriers that will exhibit high bit error probabilities in the OFDM symbol to be transmitted can be identified and excluded from data transmission, the overall BER can be improved in exchange for a slight loss of system throughput. OFDM also has some drawbacks. Because OFDM divides a given spectral allotment into many narrow subcarriers each with inherently small carrier spacing, it is sensitive to carrier frequency errors, which may be caused by Doppler shift in the channel, or by the difference between the transmitter and receiver local oscillator frequencies. This frequency offset introduces inter-carrier interference in the OFDM symbol [9,10]. We propose a differential modulation method in order to combat channel with deep fading. Bits are allocated on each subcarrier so that the overall transmit power is minimized under a fixed bit error rate (BER). This project investigates differential modulation methods for combating the effects of channel fading.

2. THE SYSTEM MODEL

Accurate channel state information (CSI) (channel coefficient), is required at the transmitter to achieve the benefits. Imperfect CSI arises from noisy channel estimates, which may also be outdated due to a delay in getting the CSI

Understanding the channel allows for manipulation of the phase and amplitude of each subcarrier at the transmitter in order to decide the appropriate modulation type for the subcarrier. A known signal is sent to the target device that enables it to build a picture of the channel environment. The target device sends back the channel characteristics to the transmitter. The transmitter can then apply the correct phase and amplitude adjustments based on the resulting SNR [11].

2.1. Impact of Channel Correlation on Channel State Information (CSI)

The performance of an OFDM system is critically dependent on the quality of independent multiple subcarriers. It is well known that channel correlation will downgrade the performance of an OFDM system, especially its capacity. Channel correlation is a measure of similarity or likeliness between the channels. In the extreme case that if the channels are fully correlated, then the OFDM system will have no difference from a single-carrier communication system.

The most general formula for calculating channel capacity in the case where channel coefficients are *either* known *or* unknown at the transmitter is the Shannon capacity formula [11]:

$$C = W \sum_{i=1}^{r} \log_2\left(1 + \frac{P_{ri}}{\sigma^2}\right) \tag{1}$$

where W is the bandwidth of each sub-channel, r is the rank of the channel coefficient matrix H. P_{ri} is the received power at each Rx antenna from the ith channel, for $i = 1, \dots, r,$ during the considered symbol time slot.

i) Unknown Channel Coefficients at the Transmitter

- **Single carrier channel**: In this case, we have $r = M = N = 1$ the channel capacity is calculated

$$C = W\log_2\left[\det\left(1 + \frac{\mathrm{P}}{\sigma^2}\right)\right] \tag{2}$$

At SNR $p = \frac{P}{\sigma^2} = 20dB,$ for instance, the normalized capacity of the single antenna channel is *C/W* = 6.658 bits/s/Hz.

- **Receive diversity**: In this case, $M = 1,\ N \geq 2$ and $\mathrm{H} = (h_1, \ldots, h_N)^T$ where $(.)^{\mathrm{T}}$ denotes the transposition operation. The channel capacity is calculated as

$$C = W\log_2\left(1 + \frac{P}{\sigma^2}\sum_{i=1}^{N}|h_i|^2\right) \tag{3}$$

Assuming that $|h_i|^2 = 1$, for $i = 1, \ldots, N,$ then we have

$$C = W\log_2\left(1 + \frac{PN}{\sigma^2}\right) \tag{4}$$

For *N = 2* and SNR *p = 2dB,* we have *C/W* = 7.6511 bits/s/Hz.

We can see that the normalized capacity in this case is larger than that in the case of channels with single Tx and Rx antennas.

- **Transmit diversity**: In this case, $M \geq 2,\ N = 1,$ and $\mathrm{H} = (h_1, \ldots, h_M)$ the channel capacity is calculated as

$$C = W\log_2\left(1 + \frac{P}{M\sigma^2}\sum_{i=1}^{M}|h_i|^2\right) \tag{5}$$

Assuming that $|h_i|^2 = 1$, for $i = 1, \ldots, M,$ then we have

$$C = W\log_2\left(1 + \frac{P}{\sigma^2}\right)$$

We see that the capacity of the channel where *channel coefficients are fixed* and *unknown at the transmitter* is the same as that of the single antenna channel regardless of the number *M* of Tx antennas.

Hence, for *M = 2, N = 1* and SNR *p = 2dB,* we have *C/W* = 6.658 bits/s/Hz.

ii) Known Channel Coefficients at the Transmitter

The channel capacity can be increased if channel coefficients are known at the transmitter. In this case, the transmitted power is assigned unequally to the Tx antennas, such that a larger power is assigned to a better sub-channel and vice versa.

- **Transmit diversity**: In this case, $M \geq 2$, $N = 1$, and $H = (h_1, \ldots, h_M)$ the channel capacity is calculated as

$$C = W\log_2\left(1 + \frac{P}{M\sigma^2}\sum_{i=1}^{M}|h_i|^2\right)$$

Assuming that $|h_i|^2 = 1$, for $i = 1, \ldots, M$, then we have

$$C = W\log_2\left(1 + \frac{PM}{\sigma^2}\right) \tag{6}$$

Hence, for $M = 2$, $N = 1$ and SNR $p = 2dB$, we have $C/W = 7.6511$bits/s/Hz which is larger than the channel capacity when the channel coefficients are *unknown* at the transmitter ($C/W = 6.658$ bits/s/Hz).

The traditional fixed resource allocation is not optimal, since the scheme is predetermined regardless of current channel conditions. On the other hand, dynamic resource allocation assigns a dimension adaptively to the users based on their channel gains. Due to the cause of time-varying nature of the wireless channel, dynamic resource allocation makes full utilization of multiuser diversity to achieve higher performance.

The block diagram of this system is shown in Figure 3. The channel estimation and mode selection are done at the receiver side and the information is sent to the transmitter using a feedback channel.

The channel estimator is used to estimate the instantaneous SNR of the received signal. Based on the instantaneous SNR calculated, the best mode will be chosen for the next transmission frame. This task is done by the mode selector block. At the transmitter the adaptive modulator block consists of different modulators which are used to provide different modulation modes. The switching between these modulators will depend on the instantaneous SNR. The goal of differential modulation is to choose the appropriate modulation mode for transmission in each sub-carrier, given the local SNR, in order to achieve good trade-off between spectral efficiency and overall BER.

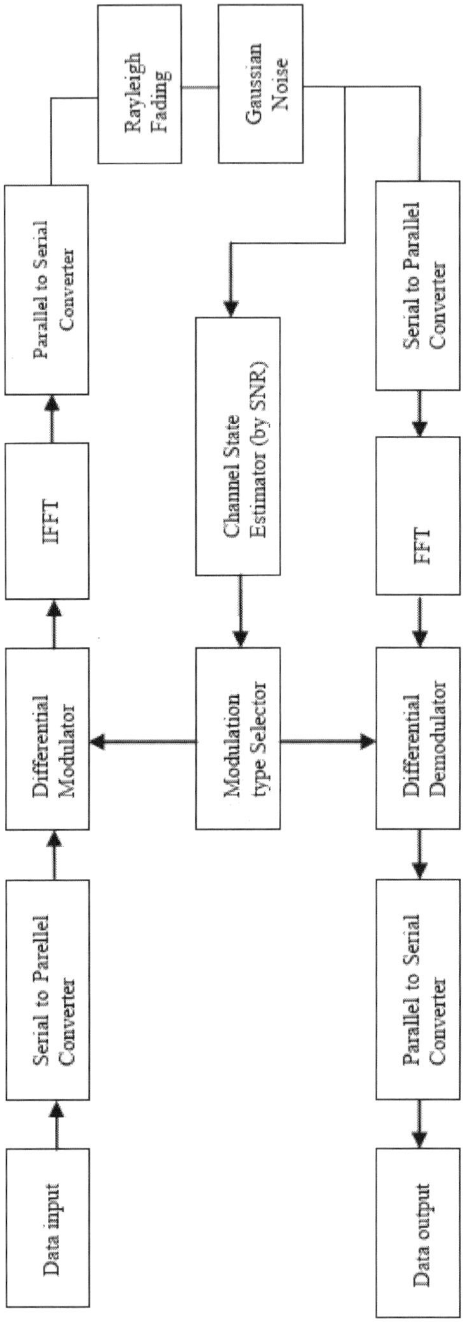

Figure 3. Block diagram of dynamic differential modulation scheme

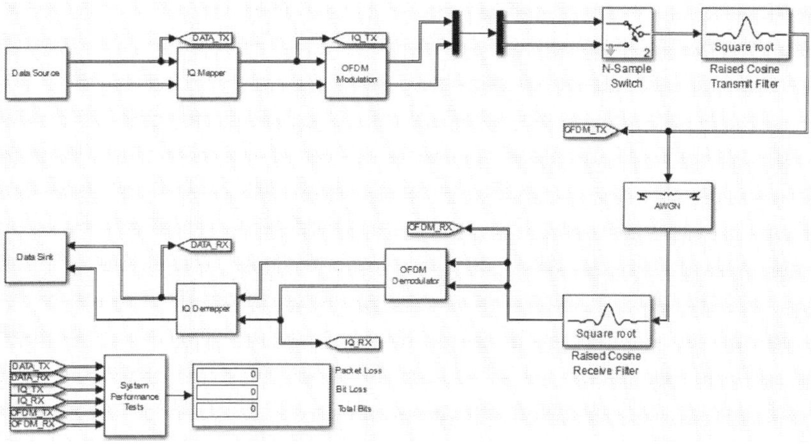

Figure 4. Simulation setup of the differential modulation scheme

The approach in setting the switching threshold is based on using the lowest quality subcarrier, i.e; the lowest value of SNR, to control the selection of modulation mode. If a suitable modulation mode or code rate is selected the overall BER can be closer to the BER target and the throughput of the system will be higher.

The switching algorithm used for the modulation schemes are presented in Tables 1 to 4. The design parameters used in the simulation are shown in Table 5.

In this modulation scheme data will be constantly transmitted even though the channel is in deep fades. If the channel quality is very bad, a robust modulation mode will be used and when the channel quality is good a spectrally efficient modulation will be used.

The increase in performance of the OFDM radio system can be measured by higher data rates, improved spectral efficiencies and the increase in channel capacity of the system.

3. SIMULATION, RESULTS, AND DISCUSSION

In this simulation, the metric for performance measurement of the differential modulation is in terms of BER. The benefits of the proposed scheme is highlighted by comparing the BER rate of the differential modulation to fixed modulation system, i.e, system with the same modulation for all the subcarriers.

Firstly, the performance of MQAM system is investigated, with the result shown in Figure 5. Then the system with all MPSK is investigated, as shown in Figures 6-10. Finally, the system a combination of MQAM and MPSK was simulated, and the results compared with the fixed modulation systems. The results are shown in Figure 11.

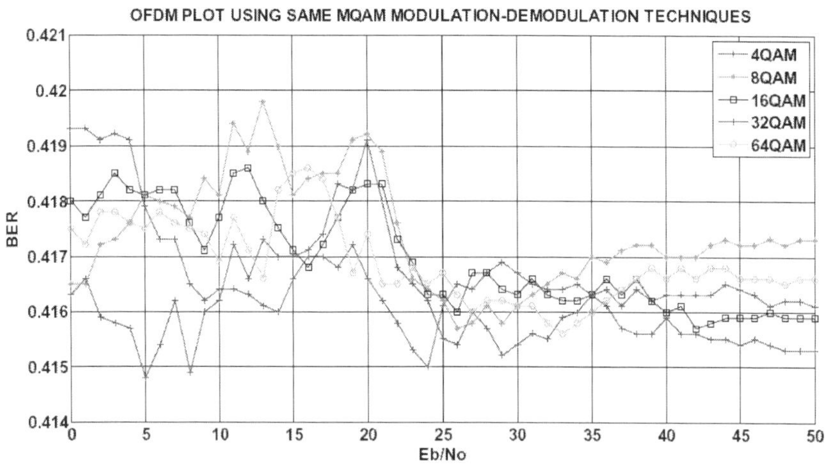

Figure 5. BER performance of fixed MQAM modulation

In the differential modulation system the BER performance was better compared to the performance of fixed modulation systems.

Figure 11 shows the performance of the differential modulation scheme. It can be seen that the BER performance of the differential scheme is at least equal to QPSK for SNR range 1 dB till 15 dB. For SNR more than 15 dB the BER performance is better than all the fixed MPSK system. However the SNR required to meet the BER target is quite high which is more than 25 dB is.

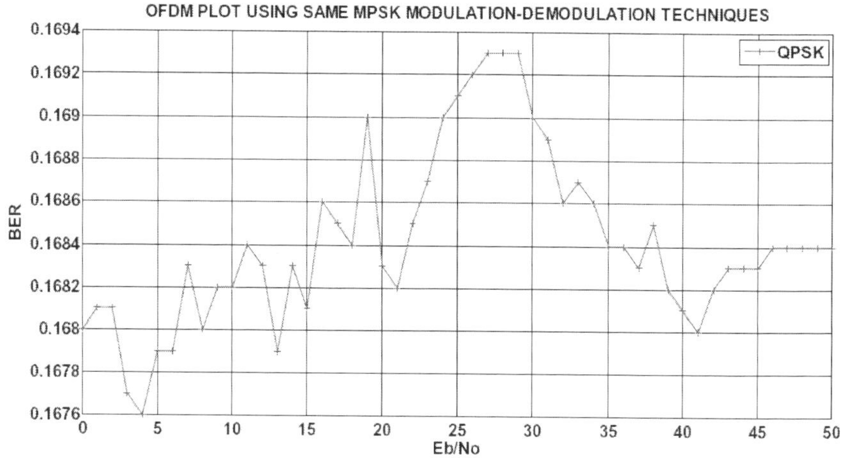

Figure 6. BER performance of 4PSK system

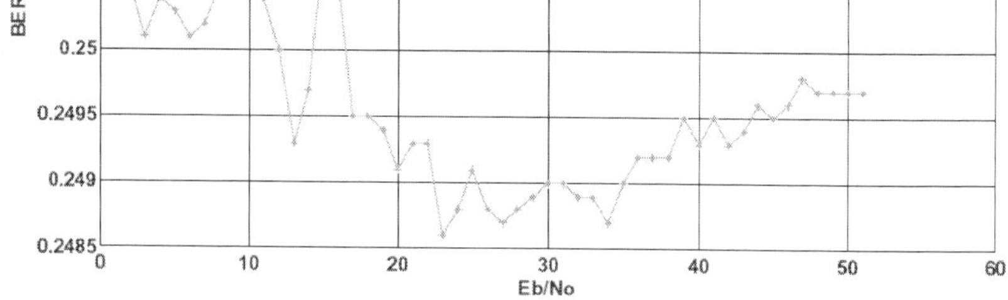

Figure 7. BER performance of 8PSK system

4. CONCLUSIONS

In this paper, the performances of adaptive transmission scheme for OFDM have been investigated. The advantage of employing adaptive transmission scheme is described by comparing their performance with fixed transmission system. A better adaptation algorithm is used to improve the throughput performance. This algorithm utilizes the average value of the instantaneous SNR of the subcarriers in the switching parameter. The results show an improved throughput performance with considerable BER performance.

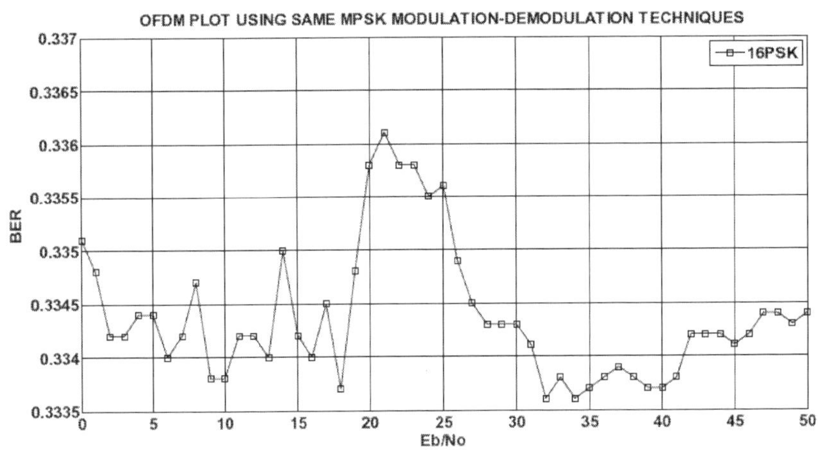

Figure 8. BER performance of 16PSK system

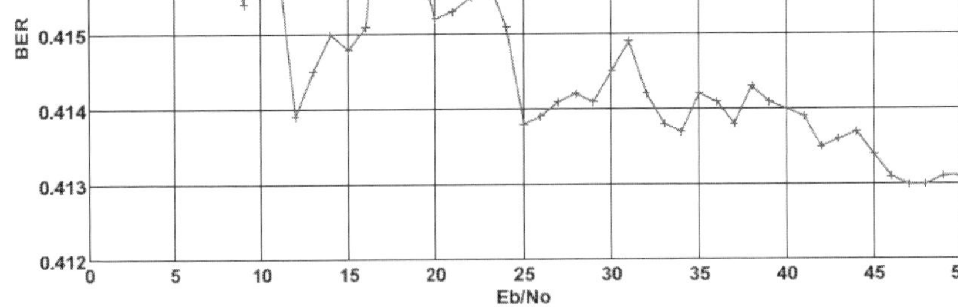

Figure 9. BER performance of 32PSK system

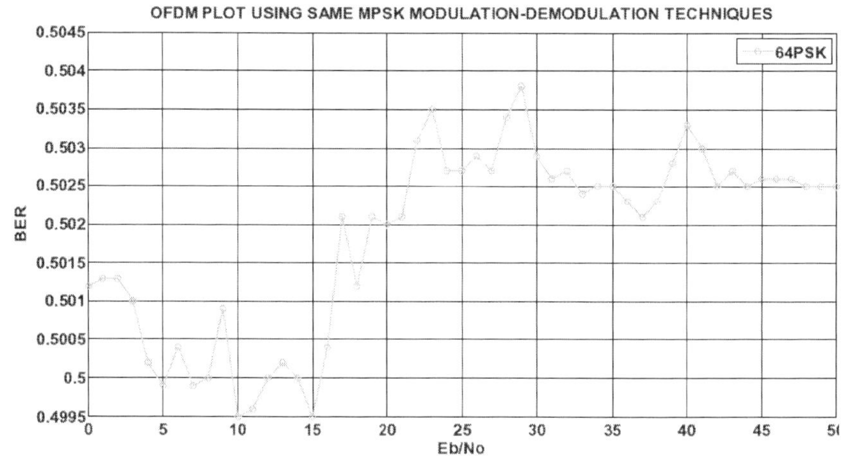

Figure 10. BER performance of 64PSK system

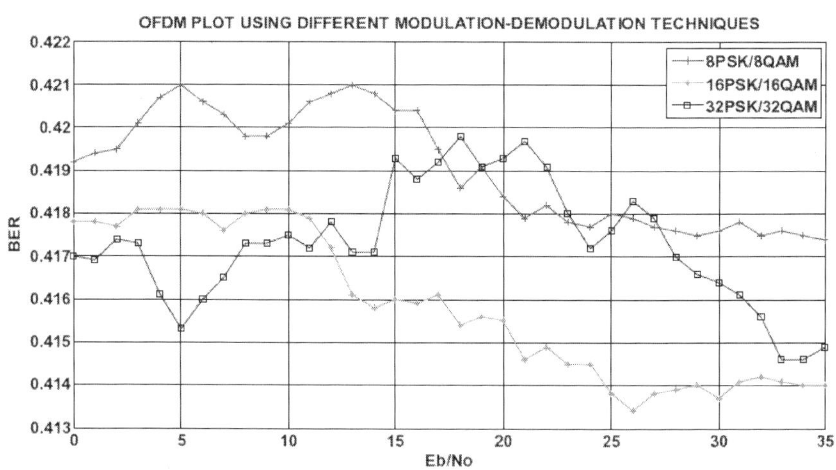

Figure 11. BER performance of the differential modulation (MQAM/MPSK)

As investigated in this paper, OFDM systems potentially possess a better BER performance when subcarriers are modulated with different modulation schemes than those with the same modulation. This benefit is described by comparing the results of the proposed scheme with the fixed scheme. In this work, this algorithm utilizes the average value of the instantaneous SNR of the subcarriers as the switching parameter. The results show an improved BER performance.

REFERENCES

1. J. Faezah and K. Sabira, 'Adaptive modulation for OFDM systems', Intl. Journal of Communication Networks and Information Security (IJCNIS) vol. 1, no.2 August 2009.

2. Sigen Ye, Rick S. Blum, Leonard J. Cimini Jr., 'Adaptive modulation for variable-rate OFDM systems with imperfect channel information.

3. Sutanu Ghosh, 'Performance Evaluation on the Basis of Bit Error Rate for different order of Modulation and different length of subchannels in OFDM system', Intl. Journal of Mobile Network Communications & Telematics (IJMNCT) vol.4, no.3, June 2014.

4. Cheong Yui Wong, Roger S. Cheng, 'Multiuser OFDM with Adaptive Subcarrier bet and power allocation', IEEE Journal on Selected Areas in Communications, vol. 17, no.10, October 1999.

5. Nitha V. Panicker, Sukesh Kumar A., 'BER Performance Evaluaation of Different Digital Modulation Schems for Biomedical Signal Transceivers, under AWGN and fading Channel conditions', Intl. Journal of Engineering and Advanced Technology (IJEAT) ISSN: 2249-8958, vol.3 iss.5, June 2014.

6. Charles U. Ndujiuba, Oluyinka Oni, Augustus E. Ibhaze, 'Comparative Analysis of Digital Modulation Techniques in LTE 4G systems', Journal of Wireless Networking and Communications, 2015, 5(2): 60-66 DOI:10.5923/j.jwnc 20150502.02.

7. M.A. Masud, M. Samsuzzaman, M.A. Rahman, 'Bit Error Rate Performance Analysis on Modulation Techniques of Wideband Code

Division Multiple Access', Journal of Telecommunications vol.1, iss 2, March 2010.

8. Umesh Sharma, 'Comparative study of Digital Modulation techniques in WIMAX', Intl. Journal of Engineering and Innovation Technology (IJEIT) vol.2, iss2, August 2012.

9. Ch. Girish Kumar, K. Sripath Roy, 'Development and Implementation of OFDM Transceiver for WLAN Applications', Intl. Journal of Engineering Research and Applications ISSN: 2248-9622, vol.4, iss.7 (version 1), July 2014, pp101-106.

10. Sangeeta Jajoria, Sajjan Singh, S.V.A.V. Prasad, 'Analysis of BER Performance of OFDM system by Adaptive Modulation', Intl. Journal of Recent Technology and Engineering (IJRTE), ISSN: 2277-3878, vol. 1, iss. 4, October 2012.

11. Charles Uzoanya Ndujiuba, Oluwadamilola Oshin, Nsikan Nkordeh, 'MIMO Deficiencies due to Antenna Coupling' Intl. Journal of Networks and Communications 2015, 5(1) 10-17,

5

INFLUENTIAL SPATIAL FACILITY PREDICTION OVER LARGE SCALE CYBER-PHYSICAL VEHICLES IN SMART CITY

Hongtao Wang, Qiang Li, Feng Yi, Zhi Li and Limin Sun

[1]Beijing Key Laboratory of IOT Information Security, Institute of Information Engineering, CAS, 100093 Beijing, China

[2]University of Chinese Academy of Sciences, 100093 Beijing, China

[3]School of Computer and Information Technology, Beijing Jiaotong University, 100044, Beijing, China

ABSTRACT

Facilities are critical infrastructures in smart city. Influence of facilities is affected by large-scale moving objects such as people and vehicles. Calculating the influence of facilities is an important task of urban computing, which adopts sensing technology to obtain objects' movement patterns in urban space and applies this information to discover many hidden issues cities face today. In this paper, we propose a computationally efficient grid partition method to compute the influence of facilities in real time, under trajectories of large-scale cyber-physical vehicles. We next predict the influence changes of facilities over dynamic vehicles using trajectory-based Markov model. We conduct evaluation using a real-world dataset, including 1-month taxi trajectories with 27,000 taxis and 1000 facilities. Experimental results shows that our solution is more efficient and achieves an accuracy of 85 %.

KEYWORDS

Smart city, Facility influence, Mobility prediction, Cyber-physical vehicle

1. INTRODUCTION

Cyber-physical vehicles [1, 2] have emerged to focus on the integration of sensing, computing, and actuation technologies into everyday urban settings and lifestyles. In cyber-physical vehicle system, locations of vehicles are the key information which not only interact dynamics of connected vehicles but also provide people a new perspective to modern city [3, 4, 5]. Understanding the influences of dynamic vehicles may help urban planners to make more informed decisions. For instance, urban planners can use massive GPS location data from taxis to help analyzing traffic anomalies [6], area functions [7], travel routes [8], etc.

Nowadays, modern urban has a variety of public infrastructures including subway stations, overpasses, office facilities, sports arenas, supermarkets, convention centers, hospitals, and entertainment parks, to name a few. They support different needs of people's urban lives and serve as a valuable tool for getting a detailed understanding of how a metropolitan area works. These infrastructures are called *facilities* [9, 10, 11], which aim to provide location-dependent services in smart city.

To monitor the service behavior of facilities, various sensor networks have been deployed in facilities. Previous sensor-based works mainly focused on coverage [12], routing [13, 14], data aggregation [15, 16], and sensory data analytics [17]. However, moving people and vehicles outside facilities also need to be monitored and analyzed. On the one hand, facilities have varying degrees of influence that heavily affects people's lifestyle. On the other hand, vehicles can generate time-varying and dynamic influence to nearby facilities. For instance, overpasses have different influences over vehicles in different times. An overpass in downtown is crowded by vehicles during rush hours on weekdays than in the suburb. In the evening, however, there are more vehicles close to overpasses in the suburb than in the downtown area. As we see from the example, dynamic mobility of vehicles significantly affects the influence of facilities. What the urban administrator cares about is what's the most influential facilities [11, 18] now and in the future. In this paper, we compute and predict the influence of facilities by considering massive dynamic moving vehicles.

While the idea of using the count of moving vehicles to compute and predict the influence of facilities seems simple, several challenges arise in practice. The first challenge is how to *efficiently* calculate the influence of facilities given massive vehicles and facilities. Urban real-time monitoring system requires facility influence to be updated as far as possible. Naive counting method and R-tree index [11] in geodatabase suffer from too much computing overhead. The second challenge is how to predict future influence of facilities when only current locations of vehicles are known. The number of vehicles around a facility at this moment cannot be used to compute the influence in the future. This is because vehicles move around and this dynamic movements leads to varying influence of facilities over time.

In this paper, we propose our ideas to address these two challenges based on the following intuitions. Firstly, some time-consuming operations such as traversing and distance computing can be replaced by lightweight mapping functions approximately. We use a grid index method which maps a vehicle location to a specific cell, and the count of vehicles within this cell is added to the influence of its nearest facility. Another intuition is that, every vehicle has its own trajectory and if two vehicles have similar trajectories, they will go to the same area in next time unit with a high probability. In particular, we use trajectory-based Markov chain model (TMM), a variation of Markov chain learned by historical movements, to predict new vehicles' movements and then update future facility influence.

We validate our solution using a large real-world dataset which consists of 1-month moving trajectories of taxis in Beijing. Facilities of interest are subway stations, overpasses, and shopping malls extracted from point of interests (POIs) in Beijing. Experimental results show that our solutions are effective and efficient in computing and in predicting facility influence under large-scale cyber-physical vehicles.

The rest of the paper is organized as follows. Section 2 presents the problem of facility influence calculation. We then propose a grid partition approach to efficiently compute facility influence in Section 3. Trajectory-based Markov model is designed to predict vehicle's location in the future (Section 4). In Section 5, we evaluate our solution using a large-scale dataset. Finally, we introduce related work and make concluding remarks.

2. PROBLEM DEFINITION

In this section, we discuss the definition of facility influence over moving objects such as vehicles and propose two problems.

Influential facility computing was first introduced in [11] where the influence of facility f is defined as the number of objects that consider f as the nearest site among all objects. A modern city has many facilities such as supermarkets, subway stations, and entertainment parks. These facilities provide a variety of services to people and cars. As a result, peoples or vehicles often gather around these facilities. When and how they go to a particular facility are determined by many factors including the services provided, the quality of the facility, and people's lifestyle and so on. No matter why objects go to different facilities, more objects means the larger influence the facility has. Figure 1 shows three facilities $\{f_1, f_2, f_3\}$ having 3, 2, and 1 vehicle(s), respectively. In this case, f_1 has the highest influence.

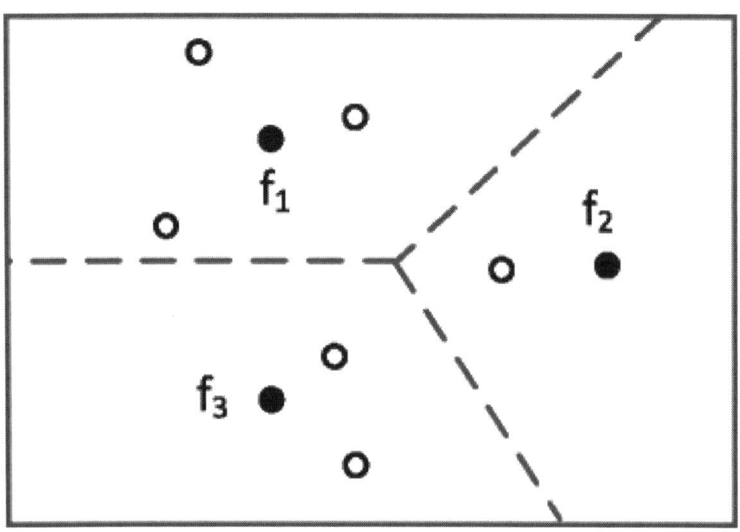

Figure 1. Facility influence. *Black dots* are facilities and *white circles* are vehicles

In general, moving objects include but are not limited to vehicles and people. The proposed algorithms in this paper can also be used for facilities and other moving objects. In the following sections, we do not differentiate terms "vehicle," "object," and "moving object". Let PP denote a set of objects and FF a set of facilities. Assume object $p \in P$ and facility $f \in F$. We denote the influence of facility f as

Definition 1 (Facility Influence).

For every facility f, the influence of f is defined as the total number of objects in its influence range at specific time stamp.

$$I_{\mathcal{P}}(f) = \sum_{p \in \mathcal{P}} R_f(p) \tag{1}$$

where $R_f(p)$ is a Boolean function: it equals 1 if object p appears in the range of the facility f, otherwise equals 0. Function $R_f(p)$ determines whether the object p is within the range of facility f. Euclidean distance $d(p, f)$ is used to measure the influence of facility impacted by the object p.

$$R_f(p) = \begin{cases} 1, & if\, d(p,f) \leq d(p,f'), f' \in \mathcal{F} \\ 0, & otherwise \end{cases} \tag{2}$$

With the definition of facility influence, we propose two problems that need to be addressed: (1) How to find out the current top k influential facilities efficiently and (2) How to predict top k influential facilities in future time under dynamic objects?

3. FAST COMPUTATION OF FACILITY INFLUENCE

In this section, we firstly explain a grid partition approach for real-time computation of facility influence and then we address two issues of this approach: ambiguous grid and grid size. Finally, we design a fast computation algorithm to calculate facility influence.

3.1 Partition of region

Given N facilities, calculating the nearest distance for one object requires N calculations. The computation overhead increases with the number of both objects and facilities. Assume we have N facilities and M objects,$\Theta(M\ N)$ distance calculations are needed in order to get influence values for all facilities. The computation overhead is exacerbated if vehicles are moving and update locations fastly. Obviously, it is unnecessary to calculate the distance for every pair of facility f and object p at every time stamp.

More specifically, we partition a geographic region into equal-size grids and provide an index for each grid. Since different facilities have different scope area, the number of grids covering different facilities varies. If a grid is inside the scope of a facility, the value that denotes the influence of this grid on the facility is stored. This value is simply the total number of objects inside this grid. With this approach, we do not need to calculate the distances to access all N facilities to discover the shortest one (whose time complexity is $\Theta(N)$); instead, mapping the index to the value incurs much less computation overhead (whose time complexity is ($\Theta(1)$).

Let us use an example (Fig. 2) to walk through this grid-based approach. Two facilities f_1 and f_2 are in the region in Fig. 2 a. Each has an associated scope and each has multiple objects around. The cell influence view (Fig. 2 b) shows that every grid cell has a nearest facility, and the objects view Fig. 2 c shows objects in every cell. We can overlay these two views and get all facilities' influence very quickly (Fig. 2 d).

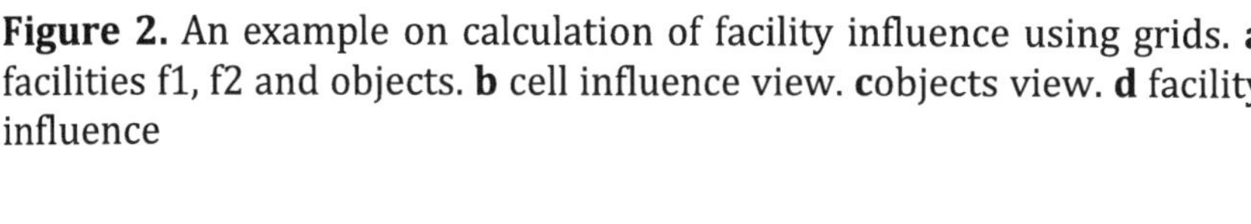

Figure 2. An example on calculation of facility influence using grids. **a** facilities f1, f2 and objects. **b** cell influence view. **c**objects view. **d** facility influence

Although the grid-based approach seems simple, two practical issues need to be addressed. First, ambiguous grid may lead to inaccuracy in calculation of facility influence: some grids are at the intersection of ranges of more than one facility. Second, the choice of the grid size significantly affects the overhead of computing facility influence. We next address these two practical issues.

3.2 Ambiguous grid

We define two types of grid cells (Fig. 3).

- *Unique grid.* A grid cell g is a unique cell if there exists a facility f, for every point pt in g such that NF(pt)=f, where NF(·) outputs the nearest facility.
- *Ambiguous grid.* A grid cell g is an ambiguous cell if there exist two different points $pt_1 \neq pt_2$ in g, such that $NF(pt_1) \neq NF(pt_2)$.

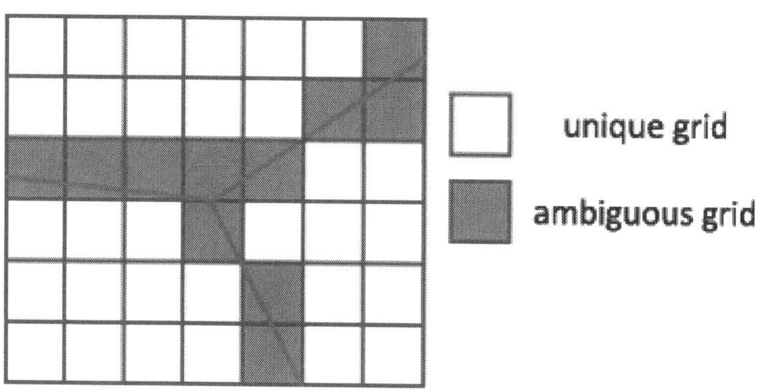

Figure 3. Unique cell and ambiguous cell

The following theorem provides a method to classify a cell being unique or ambiguous.

Theorem 1.

Let pt_i (i=1,2,3,4) be grid g's four vertices. If there exists a facility f such that for all points pt in g, $NF(pt)=f$, then$g$ is a unique cell.

Proof

Assume $p\,t_1$ is the farthest vertex from facility *f*, then for every point *pt* in *g*, *d*(*pt*,*f*)≤*d*($p\,t_1$,*f*), where *d*(*x*,*y*) is the Euclidean distance. Since *N F*($p\,t_1$)=*f*, we can get *N F*(*p t*)=*f*. Hence, *g* is a unique cell.

Due to the existence of ambiguous cells, we use *approximate influence* instead of accurate facility influence. For every facility *f* in FF, the approximate influence of facility *f* is defined as follows:

$$\hat{I}_{\mathcal{P}}(f) = \sum_{p \in \mathcal{P}} \Pr(\mathrm{NF}(g) = f | g = H(p)) \tag{3}$$

where NF(*g*) denotes the nearest facility of grid cell *g*, *H*(*p*) outputs the cell that *p* is mapped to, and Pr(·) is the probability.

Since every object can be mapped to a cell, we separate the objects set PP into two subsets: one is unique cell set P′P′ and the other is ambiguous cell set P″P″. If the mapped cell *g* is a unique cell, then Pr(NF(*g*)=*f*|*g*=*H*(*p*))=1; when the mapped cell *g* is an ambiguous cell, we design three methods to compute Pr(NF(*g*)=*f*|*g*=*H*(*p*)).

Case 1 (Centering): When object *p* is mapped to grid *g*, we use the center of *g* (denoted by *c*) representing grid*g* to compute the probability.

$$\Pr(\mathrm{NF}(g) = f | g = H(p)) = 1, \mathrm{if} \mathrm{NN}(c) = f$$

Case 2 (Sampling): We randomly choose several points in a grid and thus compute every sampling point's nearest facility in FF. The sampling points are uniformly distributed.

$$\Pr(\mathrm{NF}(g) = f | g = H(p)) = \frac{|\{s | s \in \mathcal{S}_g \wedge \mathrm{NN}(s) = f\}|}{|\mathcal{S}_g|}$$

where *s* is a sampling point and SgSg is the set of sampling points.

Case 3 (Area): We partition a grid using the Voronoi diagram, then we compute the area of each part and use the proportion related to facility *f* as the probability. Suppose all points in R_1 are in *f*, then

$$\Pr(\mathrm{NF}(g) = f \,|\, g = H(p)) = \frac{\mathrm{Area}(R_1)}{\mathrm{Area}(g)}$$

Centering is the simplest scheme as it uses center point to represent a grid and finding the nearest facility of a point is easy. However, it is too coarse-grained because it increases the inaccuracy if one object does not lie in the same region as the center. If locations mapped to a grid follow uniform distribution, sampling and area schemes are recommended. The area scheme is a generalization of the sampling scheme and can use the information of Voronoi diagram [19] generated by facility set FF.

A natural question now arises: what is the difference between a facility's actual influence and approximate influence? If we assume objects follow uniform distribution, then we have the following theorem.

Theorem 2

Given a region *R*, facility set FF, object set PP, and a grid partition with grid size *l*, for a facility f∈F, we have

$$\lim_{l \to 0} \hat{I}_{\mathcal{P}}(f) = I_{\mathcal{P}}(f)$$

Proof

Set PPconsists of two subsets: a unique cell set P′P′ and an ambiguous cell set P″P″. When *l* gets smaller, the number of objects in P′P′ decreases and thus the ambiguity of influence of facility *f* will decrease. Therefore, there exits a small *l* *, such that for all *l*<*l* *, P″=ΦP″=Φ. Thus, we have

$$\begin{aligned}\hat{I}_{\mathcal{P}}(f) &= \sum_{p \in \mathcal{P}' \cup \mathcal{P}''} \Pr(\mathrm{NN}(g) = f \,|\, g = H(p)) \\ &= \lim_{l \to 0} \sum_{p \in \mathcal{P}'} \Pr(\mathrm{NN}(g) = f \,|\, g = H(p)) \\ &= I_{\mathcal{P}}(f)\end{aligned}$$

Theorem 2 shows that if *l* is small, the approximate influence of a facility is very close to its actual influence. We can use I^P(f) to approximately estimate IP(f) if the grid size is not too big. In practical, we also shall not set the grid size too small. We demonstrate the reason below.

3.3 Grid size

Grid size l has an impact to facility influence on three aspects: memory size requirements, computing overhead, and accuracy. We next analyze these three factors and discuss how to determine a suitable grid size.

Memory Intuitively, smaller grid size makes a larger grid count in a certain geographic region. For instance, assume Beijing has an area of S (nearly to 16,800 km^2), the grid count is equal to S/l^2 where l is the grid length. If one grid only stores one value, i.e., facility influence (assume 32-bit), then the total memory needed is equal to $(S/l^2)*32$ bits. We generally do not set grid size too small considering memory usage.

Computing overhead. Given N facilities and M objects, each object p falls into one grid, so overhead of computing facility influence is $\Theta(1)$ for this object-facility pair. The total computing overhead is $\Theta(M)$, which is irrelative with grid size.

Accuracy If we set l larger, we have fewer unique cells and more ambiguous cells, which will decrease the accuracy of facility influence. The principle is not to set l too large. Empirically, we set l to be average distance of adjacent facilities. In the evaluation section, we would discuss the grid size impacting over facility influence.

3.4 Facility influence computation

After the entire geographic region is divided into cells, we can get a mapping between cells and facilities. Given an object's location, the grid index indicates which cell the object should be mapped. The corresponding nearest facility and its probability can be retrieved. Therefore, it is unnecessary to compute distance between all facilities and the object to determine the nearest facility. Our iterative process works as shown in Algorithm 1. When an object's location is updated (i.e., changed), we use grid index to get a new grid cell ID. If the current cell ID is not equal to the previous one, we extract the new nearest facility ID which is stored in the grid data structure. We then compare the current facility ID with the previous one. If they are equal, we just skip and do nothing. Otherwise, we plus facility influence value into current facility and minus it from the previous one. Then, we calculate the facility influence using (3).

Algorithm 1: Grid-based top k facility influence computation

Input: objects set $\mathcal{P}$, and facilities set $\mathcal{F}$;

Output: top k facilities set $\mathcal{F}_k$;

1 **foreach** *object p in* $\mathcal{P}$ **do**

2   Get current cell gid_c and previous cell gid_p of object p **if** $gid_c \neq gid_p$ **then**

3   Get nearest facility f_c of gid_c and nearest facility f_p of gid_p **if** $f_c \neq f_p$ **then**

4   Update f_c and f_p via (3).

5 $\mathcal{F}_k \leftarrow$ top k score of facilities

6 **return** $\mathcal{F}_k$

Analysis: The time complexity of Algorithm 1 is $O(N)$, where N is the number of objects. Lines 2 and 3 describe that the grid cell ID is obtained from object's location using grid index, so the calculation is constant time. Lines 5 and 6 retrieve data from grid data structure, which also takes constant time. The upper bound of algorithm is determined by the the number of objects.

4. INFLUENCE PREDICTION

The influence of a facility is determined by the number of vehicles gathered around it. Intuitively, vehicles have their trajectories and their future locations can be predicted from their historical movement patterns. First-order Markov chain model (MM) can be used to estimate the probability of a vehicle's next location, but it suffers from weak predictive ability. In addition, it is hard to learn a high-order Markov model due to the data sparsity problem. In this section, we present how to use TMM to predict objects' location in the future based on their trajectories. Specifically, we first discuss how to construct a Markov model, then present the details of TMM. Finally, we propose an algorithm that predicts a facility's future influence under TMM.

4.1 Construction of Markov model

We have previously used a grid partition-based approach to calculate facility influence. Now, we utilize the *same*grid representation as in the previous section to predict a vehicle's future location. Partitioned grids can be converted to a grid graph, where each node is one grid and an edge exists between two nodes implying that the two grids are adjacent.

There are two advantages using a grid partition-based approach. First, a grid representation is appropriate for facility influence prediction. Second, a grid-based approach can avoid the data sparsity problem when predicting future location. When a trajectory consists of a sequence of locations, it can be represented using a sequence of grids. Therefore, we will have lots of trajectory data in grid index to represent each trajectory. For predicting future locations with historical trajectories, lots of training data is needed for our TMM model.

Each grid cell is viewed as a state and the Markov model uses these states to build its state transition matrix. In the matrix, the transition probability between grid cells (i.e., states) is trained by the historical step counts, i.e., the number of transition steps from one cell to another is used to derive the transition probability between these two cells. To further understand the processing of Markov model for trajectories, we use an example to illustrate it. In Fig. 4, the region is partitioned into 3×3 grids. Trajectory T_1 is a sequence of grid cells: g_1, g_4, g_5 (Fig. 4 a). Each grid is a state in Markov model, so T_1 is represented as a sequence of states over time (Fig. 4 b).

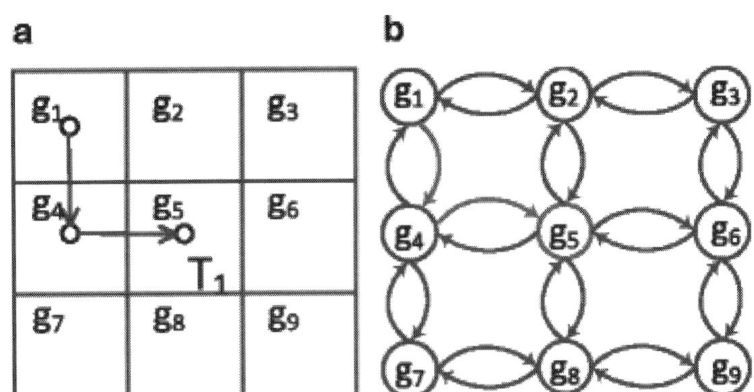

Figure 4. An example of using Markov model. **a** grid partition for trajectory. **b** 3×3 Markov model for trajectory

The prediction of objects' future locations consists of two phases: a training phase where the historical trajectories are mined offline to obtain transition matrix and an online prediction phase where the historical trajectory of a given object is analyzed and its future locations are predicted.

In the training phase, a Markov model is established by associating a state with each grid g_i. Two directed transitions of states corresponding to adjacent grids g_i and g_j are established, i.e., g_i to g_j and g_j to g_i. Let S_{ij} denotes a travel from g_i to g_j. The transition probability of S_{ij} is denoted by $Pr(S_{ij})$. Note that a trajectory T_{ij} is composed of a set of steps: $T_{ij}=\{S_{i,i+1},S_{i+1,i+2},\cdots,S_{i+n,j}\}$. Formally,

$$Pr(S_{ij}) = \frac{|S_{ij}|}{|S_{i\neg}|} \tag{4}$$

where $|S_{i\neg}|$ is the total number of travels from g_i to all neighboring cells.

For each pair of adjacent nodes in the grid graph, we compute the transition probabilities offline using Eq. (4). These probabilities are stored as entries of a two-dimensional matrix where one dimension corresponds to the node of current state and the other dimension corresponds to the next state. We denote the matrix and its entries by the transition matrix M and M_{ij}, respectively.

4.2 Prediction of future locations

Prediction of object's future locations essentially provides a set of cells that the object will reside along with a probability. We next use an example (Fig. 5) to illustrate how to predict future location using TMM model. Suppose one object p currently stays at grid cell g_5. Next time, it has four possible movements: g_2, g_4, g_6, andg_8 (Fig. 5 a). If its trajectory is unknown, the transition probability (in the transition probability matrix of Markov model) is directly used as the probability of it going to each of the four cells. If its trajectory is known, we can use Bayesian theorem to calculate the probabilities (Fig. 5 b), where prior probability is the transition probability and the likelihood probability is how the trajectory and the historical data is similar to each other.

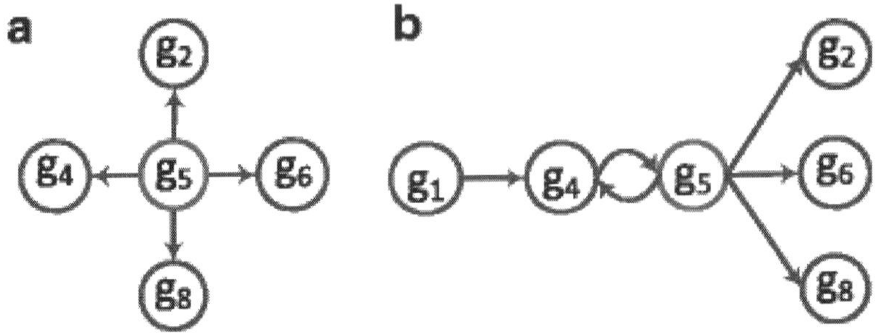

Figure 5. Trajectory-based Markov model. **a** Normal transition. **b** Transition based on trajectory

We then formulate the prediction with objects trajectory based on TMM. Assume one object is currently in cell g_j with trajectory T_{ij}, i.e., it has gone through a certain path from g_i to g_j. Now the question is how to derive the probability of the object arriving at an adjacent cell g_k in the next time unit, given the knowledge of T_{ij}? This probability can be computed using Bayes rule as follows.

$$Pr(S_{jk}|T_{ij}) = \frac{Pr(T_{ij}|S_{jk})Pr(S_{jk})}{Pr(T_{ij})} \tag{5}$$

where $Pr(S_{jk})=Pr_{jk}$ can be obtained from transition matrix M. $Pr(T_{ij}|S_{jk})$ denotes the probability of an object traveling from g_i to g_j given that transition steps are from grid j to grid k. It can be computed as follows [20].

$$Pr(T_{ij}|S_{jk}) = \frac{Pr(T_{ij}) \cdot Pr(S_{jk})}{Pr_{i \to k}} \tag{6}$$

where $Pr_{i \to k}$ is the total transition probability of all possible trajectories traveling from g_i to g_k. Combining Eq. (6) and Eq. (5), we have

$$Pr(S_{jk}|T_{ij}) = \frac{Pr^2(S_{jk})}{Pr_{i \to k}} \tag{7}$$

$Pr(S_{jk})$ can be obtained from transition matrix M. In fact, $Pr_{i \to k}$ can also be calculated from M. In general, $M^r (r \in [0,\infty))$ holds the probabilities of transition from one node to another in exactly r steps (i.e., M^r holds r-step transition probabilities). One method [20] to compute $Pr_{i\to k}$ is

$$Pr_{i\to k} = M_{ik}^{L_{i\to k}} \cdot \left(M_{ik}^{0} + M_{ik}^{1} + \cdots + M_{ik}^{L_{de,i\to k}}\right) \tag{8}$$

where $L_{i\to k}$ denotes l_1 distance (step count of the shortest path) from g_i to g_k and $L_{de,i\to k}$ denotes the maximal steps where an object may run more paths from g_i to g_k. $L_{de,i\to k}$ is often set to be $0.2L_{i\to k}$.

4.3 Prediction of future facility influence

Facility influence is determined by objects around them. If we can predict these objects' future locations, then we can predict facility influences in the future. This raises several interesting challenges. One challenge is how to select a proper time step size. Objects have different speed and direction, e.g., taking a taxi requires a larger time step size compared with walking. Mobility is also affected by surrounding environments, such as traffic congestion or speed limitation. It is hard to predict when objects will arrive at the predicted place even though we make a successfully prediction. The other challenge is that the prediction is related to the size of grid cell. If the cell is too small, object may cross several cells in one time step, resulting in data sparsity problem for the training phase of the Markov model construction; if the cell is too large, object will stay in one cell for several time steps, resulting in inaccuracy to facility influence. In our experiments, we have tried multiple cell size from 100 to 1000 m and empirically selected an appropriate value.

The speed of an object is related to the object's previous speed. If we can get these two pieces of information as well as the speed, we can determine the grid cells the object will arrive. To be specific, suppose travel time is t, the speed is v, so the distance from current location is s=v t. Then, we can get a set of possible cells C within distance s from current location. If cell g in C is next to current cell, we can directly use Eq. (7) to compute the probability. After getting all possible cells' probability, we can use the highest one as the final prediction result for that object. Algorithm 2 summarizes how prediction of facility influence is done.

Algorithm 2: Facility influence prediction

Input: updated trajectory set $\mathcal{T}$ and grid transition Matrix M;
Output: Top k facilities' set $\mathcal{F}_k$ in next time slot;

```
foreach trajectory {T_ij} do
    Evaluate all possible cell set C;
    foreach every cell in C do
        Compute Pr(S_jk | T_ij) using M and Eq. (7);
    Identify the cell g with maximal probability;
    Compute the influence of the facility in grid g;
{f} ← Top k score of facilities;
return {f}
```

---

5. PERFORMANCE EVALUATION

Our evaluation consists of three parts: (1) memory usage and computing overhead for grid partition and indexing, (2) overhead and accuracy of facility influence computing, and (3) prediction accuracy of facility influence prediction.

5.1 Experimental methodology

To validate the effectiveness and efficiency of proposed algorithms, we take taxis as moving objects for facility influence computation. We use a large real-world dataset which consists of 1-month moving trajectories of taxis in Beijing. It contains more than 27,000 taxis over one billion trajectories. Facilities are overpasses, subway stations, tourism destinations, shopping malls, etc, which are extracted from over 1000 Beijing POIs [7]. Each taxi uploaded its location to the database server every 30 s. Figure 6 a shows the coverage of taxis in 1 day. We can see that taxi trajectories almost cover all central urban area of Beijing. Figure 6 b is a typical taxi's trajectory in a day.

We ran our proposed algorithm on a desktop machine with Intel Core I5-3380 2.90GHz dual core CPU. The data is stored in SQL server database. There are many data structures to construct grid index, such as hash table, hash array, or matrix. We use hash array to index the grids. Each grid use 100 bits to store two values: facility IDs and the corresponding influence value. We train our trajectory-based Markov model with built-in toolkit [21]. We first separate every taxi trajectory into sub-trajectories which starts in one grid cell and ends in another. Then, we split the sub-trajectory database to two parts, one for training Markov transition matrix and the other for testing the effectiveness of our algorithm. We then compute and save the Markov transition matrix M using Eq. (4). It takes several hours to train matrix M using the massive trajectories. We then compute $M^2, M^3, \cdots$ and save them on a hard disk. When doing online prediction, we put all the matrices into memory, avoiding slow matrix multiplication operations.

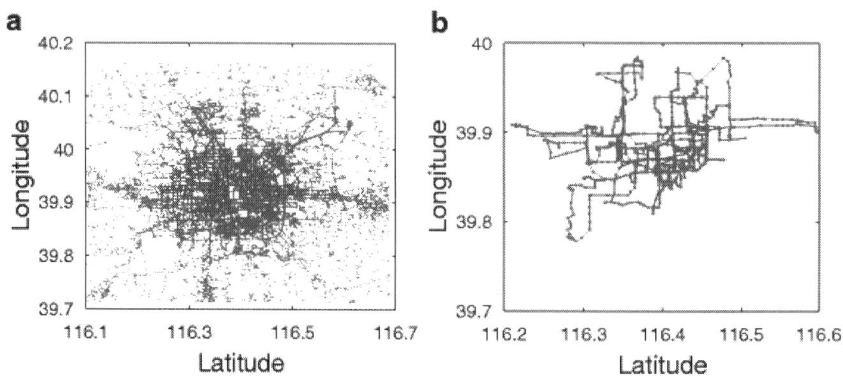

Figure 6. Distribution of taxi trajectories. **a** Trajectory coverage. **b** A taxi's trajectory 1 day

For evaluation of computing overhead, we use response time to measure the efficiency of our algorithms, which includes CPU time and I/O latency. For evaluating accuracy, we compute two type of positive cases: correct influence calculation ratio and correct influence prediction ratio. Table 1 lists all parameters used in our experiments.

5.2 Grid partition

We first measure memory usage and overhead of grid partition. The memory usage is on the grid index. Table 2lists memory usage of grid index with various grid sizes, increasing from 10 to 1000 m; 10 m is close to GPS location precision. The larger grid size, the more memory space in a fixed geographic region. In our experiments, Beijing has an area size of 16,800 km², with the smallest grid size of 10 m (GPS precision is close to 10 meters), we need 2.5 GB memory. This is acceptable in our case.

Table 1. Notations

Notation	Definition	Value
n	Taxi count	27,000
m	Facility count	300 to 1000
k	The most top-k influential facilities	10 to 100
l	Grid cell size	100 to 1000 m

Table 2. Grid index memory usage

l	10	100	200	500	1000
Grid index size(MB)	2500	25	7.5	1	0.25

Figure 7 depicts the running time to different grid partitions. The running time increases along with the increase in the number of facilities from 100 to 500, in Fig. 7 a. This is because grid partition needs to calculate every facility's scope and then map it into grid cells, larger facility count leads to higher computing overhead. Figure 7 bdepicts the impact of grid size on running time. It confirms our intuition that smaller grid size causes higher computing overhead in grid partition. The smaller the grid size, the higher the grid count in a fixed geographic region.

In brief, if grid size *l* is too small, the index requires larger memory space and higher overhead. We propose to set *l* from 100 to 500. The facility count depends on specific application scene, which affects the computing overhead.

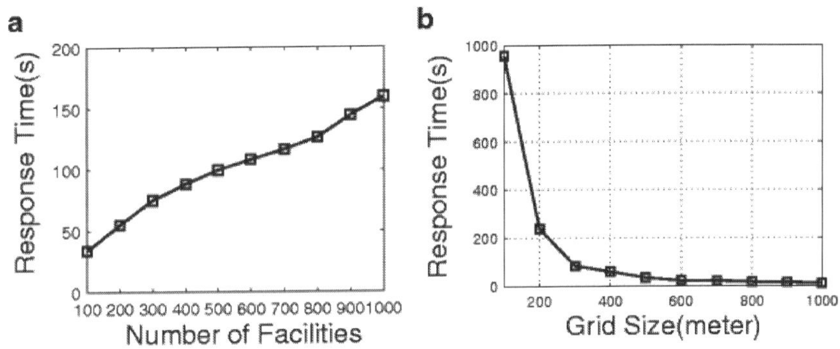

Figure 7. Computing overhead of grid partition. **a** Impact of facility count on overhead in grid partition. **b** Impact of grid size on overhead in grid partition

5.3 Facility influence computing

We evaluate Algorithm 1 for calculating the facility influence. Figure 8 a depicts how taxi count affects the response time of facility influence calculation. In these experiments, we set grid size l=100 and facility count$m$=300. The calculation only takes 100 ms when the taxi count is close to 10,000. The response time increases very slowly with the increase of taxi count n. This is because the calculation maps location to grid index, which has a constant complexity $\Theta(1)$. Figure 8 b shows how facility count affects response time. In these experiments, grid size is l=100 and taxi count n=10,000. Our approach remains very efficient and the response time is under 0.5 s. These results have confirmed that using grid is suitable for calculating facility influence in real time.

We next evaluate the effectiveness of dealing with ambiguous grids. We basically measure the accuracy of associating objects with facility. We consider the ground truth being the closest facility in Euclidean distance. In these experiments, facility count m=300 and taxi count n=10,000. Figure 9 depicts the results using the three methods we proposed: *centering* (a), *sampling* (b), and *area* (c). If grid size is too large, there are large ambiguous grids at the intersection of multiple facilities. The accuracy decreases with the increase of grid size. When the grid length becomes 1000 m, the accuracy of all three methods is unacceptable with the worst case close to 56 %. When the grid size is small (i.e., close to 100 meters), the accuracy of all three methods stays high (i.e., close to 96 %.) Combining the results of grid partition shown in Fig. 7 and Table 2, large grid size indicates low memory usage and low overhead. We suggest to set *l* between 100 and 500 m according to different surroundings (various facility types). Fig. 9 also depicts the accuracy with different *k* values, i.e., calculating top *k*most influential facilities. If *k* equals facility count *m*, it basically calculates influence for all facilities. Here, we measure different *k* values impacting facility influence calculation (total count is 300), including 10, 50, 80, and 100. There is no difference in accuracy regardless of the *k* value.

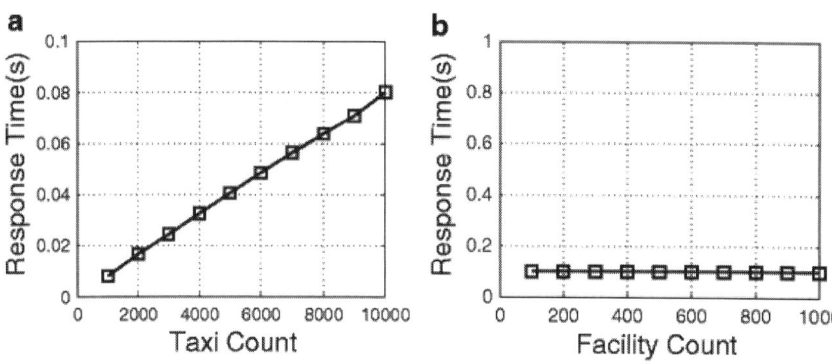

Figure 8. Overhead of calculating facility influence. **a** Impact of taxi count on overhead of facility influence calculation. **b**Impact of facility count on overhead of facility influence calculation

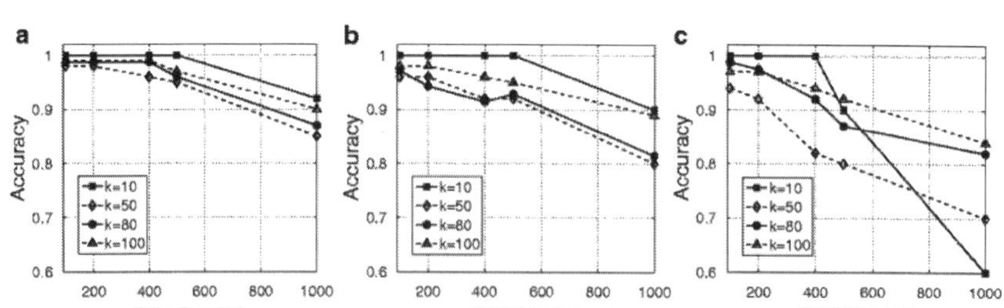

Figure 9. Accuracy of associating objects with facility. **a** Centering. **b** Sampling. **c** Area

5.4 Prediction accuracy

We firstly evaluate location prediction accuracy of our TMM. Two factors affect the accuracy: the trajectory length and the number of steps taken. Step indicates that the location is transformed from one grid cell to another cell (includes jumping circle). Figure 10 depicts the impact of trajectory length on location prediction accuracy (assume future step is 1). When the trajectory length is long (10 steps in this case), the accuracy is close to 85 %. When the number of historical steps equals 1, the accuracy is below 70 %. This is reasonable since more historical data will lead to higher accuracy in prediction. The cumulative distribution function (CDF), as shown Fig.10 b, depicts that the accuracy is higher than 80 % on average.

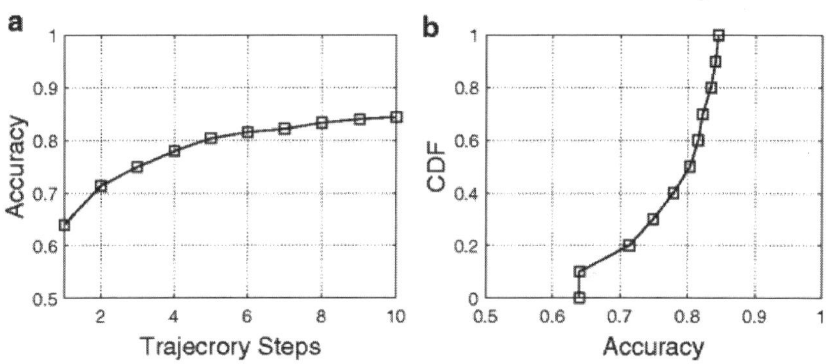

Figure 10. Prediction accuracy with trajectory length. **a** Historical step count. **b** CDF under historical steps

In order to verify the effectiveness of our TMM for future step count, we compare it with the naive MM which is considered as baseline. MM is a common methods in location prediction with normal transition, which directly uses the transition probability of current state to neighbors. The trajectory length used for TMM and MM ranges from 3 to 10.

Figure 11 a depicts impact of future steps on the prediction accuracy. The accuracy of TMM is much higher than MM in every steps, which indicates that historical trajectories can to a large extent dominate future locations. Figure 11 b shows the CDF of prediction accuracy of the two models: for 80 % of the time, the prediction accuracy in TMM is lower than 80 %. While the prediction accuracy in MM is lower than 40 %, which is significantly lower than TMM.

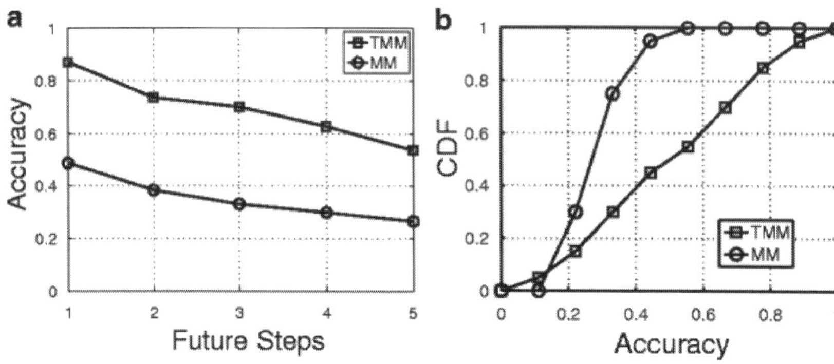

Figure 11. Prediction accuracy with future step count. **a** Future step count. **b** CDF under future steps

For the trajectory length, we suggest to use a longer one so that the prediction accuracy is higher. For large future step count, the prediction accuracy seems low from our experiments. However, it does not affect the facility influence prediction. This is because when the taxi is moving within the scope of a facility, prediction error in future location cannot cause the error in facility influence prediction. The error in influence prediction only occurs when the taxi moves from the scope of one facility to another, which case error occasionally occurs in future step prediction.

We next evaluate the effectiveness of predicting facility influence. We set the parameters l=500,m=500,n=30,000; the historical steps is bigger than 4 and we only consider the situation in 5 future steps. The ground-truth of our test uses nearest distance to calculate the facility influence. We use the *sampling* method to deal with ambiguous grids. Figure 12 a depicts the accuracy of dealing with ambiguous grids and also future influence prediction. We can see that our facility influence prediction accuracy is close to the accuracy of location prediction, i.e., above 80% in all k values tested. Our facility influence calculation accuracy has better accuracy than prediction by nearly more than 10 %, and this difference is caused by the error in future location prediction. Figure 12 b shows the CDF of facility influence prediction accuracy. Our solution has achieved an acceptable accuracy for predicting facility influence.

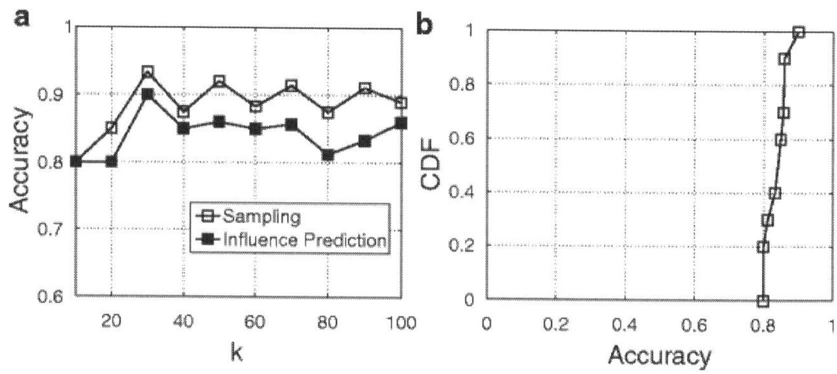

Figure 12. Accuracy of facility influence prediction. **a** Accuracy of prediction and computing. **b** CDF under prediction

We observe that although location prediction can introduce uncertainty which then affects the final influence prediction, as long as the predicted grid is not far away from the real location and they have the same nearest facility, this uncertainty has weak impact on influence prediction.

6. RELATED WORK

Our work lies at the intersection of urban computing and facility influence query in databases.

Urban computing. Urban computing is emerging as a concept where every sensor, device, person, vehicle, building, street, and so on in urban areas can be used as a component to probe city dynamics and further enable a city-wide computing for serving people and their cities. Previous work has mainly focused on solving specific issues that urban residents face, such as measuring air pollution of POIs using human mobility in different city regions [7], recommending taxi routes [22, 23], or inferring energy consumption of city via the car refueling behavior [24]. These works mostly focus on dynamic locations due to human mobility, but less on facility influence.

Cyber-physical vehicles. In general, cyber-physical vehicles are equipped with vehicular cyber-physical systems in which various vehicles are networked to sense, monitor, control, and communicate with the physical world. Cyber-physical vehicles have attracted significant attention recently [25]. Different from the existing work on cyber-physical systems, recent studies on cyber-physical vehicles focus on message dissemination optimization [26], traffic volume measurement [27], data dissemination [28], integrating with mobile cloud computing [29], and so on. Our work use cyber-physical vehicles (typically considered as dynamic objects in urban computing) to research the changes in influence of facility (usually focus on static facility in the databases community). The increasing availability of GPS-embedded cyber-physical systems provides large scale taxi trajectories in urban computing. Many works have studied taxi trajectories and then proposed strategies for improving taxi drivers' income by analyzing the pick-up and drop-off behavior of taxicabs in different locations [30,31]. We also use taxi trajectories; however, we use them to consider how vehicles impact facility influence.

Facility influence. In the database community, facility influence calculation is mainly focused on spatial query processing over different geographic regions. Korn and Muthukrishnan [9] propose reverse nearest neighbor as influence indicator to measure top k most influencial web sets. Other work [10, 11, 32] extends web set queries to the spatial place in geographic region. These previous works usually construct two R-tree structures in memory, in which objects are stationary, but writing to and reading from the database to load the memory is major cost. Different from these existing works, our work focuses on the dynamic objects in a geographic area. Several previous works [19, 33] propose the preprocessing stage to efficiently compute rather than using R-tree. Voronoi-based partition [19] divides scopes of facility into various region-irregular areas, and each area represents the scope of a facility. This approach requires to build tree index for Voronoi partition and time cost over query Voronoi index is related to area count. Reference [33] adopts adaptive grid partition to find influential locations. Similar to Voronoi-based partition, adaptive grid also requires R-tree indices to identify right grid number to calculate the facility influence. In contrast, our work adopts grid partition with uniform grids. Regular area partition can be indexed with a simple data structure, so its query time is constant.

7. CONCLUSIONS

In this paper, we propose to predict the influence of a facility in smart city, under large-scale dynamic cyber-physical vehicles. A facility supports public services to satisfy the needs of people's urban lives, and its influence is affected by the count of vehicles gathered around it. We present a grid partition approach to efficiently compute facility influence over dynamic vehicles. To predict facility influence in the future, TMM model trained by historical trajectories is proposed to predict vehicle's future movement, which is then used to evaluate future influence of facilities. Experimental results show that our approaches are acceptable in terms of both computing facility influence in real time and predicting future facility influence with high accuracy.

ACKNOWLEDGEMENTS

This work was supported by the National Natural Science Foundation of China (Grant No.61402463), the "Strategic Priority Research Program" of the Chinese Academy of Sciences (Grant No.XDA06040101), and the Young Talent Program of Institute of Information Engineering Chinese Academy of Sciences (Grant No.1102008202).

REFERENCES

1. M Lukasiewycz, S Steinhorst, F Sagstetter, W Chang, P Waszecki, M Kauer, S Chakraborty, in *15th Euromicro Conference on Digital System Design, DSD 2012*. Cyber-physical systems design for electric vehicles (Cesme, IzmirTurkey, 2012), pp. 477–484.

2. DPF Möller, T Deriyenko, H Vakilzadian, in *IEEE International Conference on Electro/Information Technology, EIT 2015*. Cyber-physical vehicle tracking system: Requirements for using a radio frequency identification technique (Dekalb, ILUSA, 2015), pp. 552–557.

3. T Kindberg, M Chalmers, E Paulos, Guest editors' introduction: Urban computing. Pervasive Comput. IEEE. **6**(3), 18–20 (2007).

4. Y Zheng, L Capra, O Wolfson, H Yang, Urban computing: Concepts, methodologies, and applications. ACM Transaction on Intelligent Systems and Technology. **5**(3), 222–235 (2014).

5. Y Zheng, Y Liu, J Yuan, X Xie, in *UbiComp 2011: Ubiquitous Computing, 13th International Conference*. Urban computing with taxicabs (UbiComp 2011Beijing, China, 2011), pp. 89–98.

6. S Chawla, Y Zheng, J Hu, in *12th IEEE International Conference on Data Mining, ICDM 2012*. Inferring the root cause in road traffic anomalies (BrusselsBelgium, 2012), pp. 141–150.

7. J Yuan, Y Zheng, X Xie, in *The 18th ACM SIGKDD International Conference on Knowledge Discovery and Data Mining, KDD '12*. Discovering regions of different functions in a city using human mobility and POIs (ACMBeijing, China, 2012), pp. 186–194.

8. L Wei, W Peng, W Lee, Exploring pattern-aware travel routes for trajectory search. ACM TIST. **4**(3), 48 (2013).

9. F Korn, S Muthukrishnan, in *Proceedings of the 2000 ACM SIGMOD International Conference on Management of Data*. Influence sets based on reverse nearest neighbor queries (Dallas, Texas, USA., 2000), pp. 201–212.

10. J Huang, Z Wen, J Qi, R Zhang, J Chen, Z He, in *Proceedings of the 20th ACM Conference on Information and Knowledge Management, CIKM 2011*. Top-k most influential locations selection (Glasgow, United Kingdom, 2011), pp. 2377–2380.

11. T Xia, D Zhang, E Kanoulas, Y Du, in *Proceedings of the 31st International Conference on Very Large Data Bases*. On computing top-t most influential spatial sites (Trondheim, Norway, 2005), pp. 946–957.

12. J Gao, J Li, Z Cai, H Gao, in *2015 IEEE Conference on Computer Communications, INFOCOM 2015*. Composite event coverage in wireless sensor networks with heterogeneous sensors (KowloonHong Kong, 2015), pp. 217–225.

13. Z Cai, Z Chen, G Lin, A 3.4713-approximation algorithm for the capacitated multicast tree routing problem. Theor. Comput. Sci. **410**(52), 5415–5424 (2009).

14. Z Cai, G Lin, G Xue, in *Computing and Combinatorics, 11th Annual International Conference, COCOON 2005*. Improved approximation algorithms for the capacitated multicast routing problem (KunmingChina, 2005), pp. 136–145.

15. Z He, Z Cai, S Cheng, X Wang, Approximate aggregation for tracking quantiles and range countings in wireless sensor networks. Theor. Comput. Sci. **607:**, 381–390 (2015).

16. J Li, S Cheng, H Gao, Z Cai, Approximate physical world reconstruction algorithms in sensor networks. IEEE Trans. Parallel Distrib. Syst. **25**(12), 3099–3110 (2014).

17. S Cheng, Z Cai, J Li, X Fang, in *2015 IEEE Conference on Computer Communications, INFOCOM 2015*. Drawing dominant dataset from big sensory data in wireless sensor networks (KowloonHong Kong, 2015), pp. 531–539.

18. L Zhan, Y Zhang, W Zhang, X Lin, in *21st ACM International Conference on Information and Knowledge Management, CIKM'12*. Finding top k most influential spatial facilities over uncertain objects (Maui, HIUSA, 2012), pp. 922–931.

19. P Zhang, R Cheng, N Mamoulis, M Renz, A Züfle, Y Tang, T Emrich, in *29th IEEE International Conference on Data Engineering, ICDE 2013*. Voronoi-based nearest neighbor search for multi-dimensional uncertain databases (BrisbaneAustralia, 2013), pp. 158–169.

20. AY Xue, R Zhang, Y Zheng, X Xie, J Huang, Z Xu, in *29th IEEE International Conference on Data Engineering, ICDE, 2013*. Destination prediction by sub-trajectory synthesis and privacy protection against such prediction (BrisbaneAustralia, 2013), pp. 254–265.

21. Hidden Markov Model (HMM) Toolbox. https://www.cs.ubc.ca/~murphyk/Software/HMM/hmm.html.

22. L-Y Wei, Y Zheng, W-C Peng, in *The 18th ACM SIGKDD International Conference on Knowledge Discovery and Data Mining, KDD '12*. Constructing popular routes from uncertain trajectories (BeijingChina, 2012), pp. 195–203.

23. A Bassoli, J Brewer, K Martin, P Dourish, S Mainwaring, Underground aesthetics: Rethinking urban computing. Pervasive Comput. IEEE. **6**(3), 39–45 (2007).

24. F Zhang, D Wilkie, Y Zheng, X Xie, in *The 2013 ACM International Joint Conference on Pervasive and Ubiquitous Computing, UbiComp '13, Zurich*. Sensing the pulse of urban refueling behavior (ACMZurich, Switzerland, 2013), pp. 13–22.

25. D Jia, K Lu, J Wang, X Zhang, X Shen, A survey on platoon-based vehicular cyber-physical systems. IEEE Commun. Surv. Tutor. **18**(1), 263–284 (2016).

26. H Zheng, J Wu, in *35th IEEE International Conference on Distributed Computing Systems, ICDCS 2015*. Optimizing roadside advertisement dissemination in vehicular cyber-physical systems (Columbus, OHUSA, 2015), pp. 41–50.

27. Y Zhou, S Chen, Z Mo, Q Xiao, in *35th IEEE International Conference on Distributed Computing Systems, ICDCS 2015*. Point-to-point traffic volume measurement through variable-length bit array masking in vehicular cyber-physical systems (Columbus, OHUSA, 2015), pp. 51–60.

28. K Liu, VCS Lee, JK Ng, J Chen, SH Son, Temporal data dissemination in vehicular cyber-physical systems. IEEE Trans. Intell. Trans. Syst. **15**(6), 2419–2431 (2014).

29. J Wan, D Zhang, Y Sun, K Lin, C Zou, H Cai, VCMIA: A novel architecture for integrating vehicular cyber-physical systems and mobile cloud computing. MONET. **19**(2), 153–160 (2014).

30. Y Ge, C Liu, H Xiong, J Chen, in *Proceedings of the 17th ACM SIGKDD International Conference on Knowledge Discovery and Data Mining*. A taxi business intelligence system (ACMSan Diego, CA, USA, 2011), pp. 735–738.

31. J Yuan, Y Zheng, L Zhang, X Xie, G Sun, in *UbiComp 2011: Ubiquitous Computing, 13th International Conference*. Where to find my next passenger (UbiComp 2011Beijing, China, 2011), pp. 109–118.

32. J Liu, G Yu, H Sun, in *Proceedings of the 20th ACM Conference on Information and Knowledge Management, CIKM 2011*. Subject-

oriented top-k hot region queries in spatial dataset (GlasgowUnited Kingdom, 2011), pp. 2409–2412.

33. D Yan, RC-W Wong, W Ng, in *Proceedings of the 20th ACM Conference on Information and Knowledge Management, CIKM 2011*. Efficient methods for finding influential locations with adaptive grids (Glasgow, United Kingdom, 2011), pp. 1475–1484.

6

A COGNITIVE GLOBAL CLOCK SYNCHRONIZATION METHOD IN WIRELESS SENSOR NETWORKS

Bilal Ahmad, Ma Shiwei, Lin Lin, Song Yang

School of Mechatronics Engineering and Automation, Shanghai Key Laboratory of Power Station Automation Technology, Shanghai University, Shanghai

ABSTRACT

Clock synchronization in sensor networks quite recently attained great importance and researchers are concentrating their attention on it. It is utmost difficult to get clock synchronization through clock skew and phase offset. Every sensor node has its own internal clock and it is very necessary to get all the clocks of sensor nodes precisely on common time value to achieve good synchronization. In surveillance and law enforcement applications global synchronization is important as precise mapping is required for the collected sensor data. The sensor nodes form a multi-hop and ad-hoc network for such applications. The basic demand for these applications is high accuracy and reduced latency error. The main contribution of this article is: 1. Brief introduction of fault tolerant novelty of the cognitive global clock synchronization algorithm (CGCSA) and 2. In this article we have observed jointly maximum likelihood estimation (JMLE) of the clock skew and clock offset in CGCSA to check its performance in terms of synchronization accuracy and compared our scheme with receiver only synchronization (ROS) strategy.

KEYWORDS

Sensor networks, Clock Synchronization, Clock offset, Clock Skew, Algorithm

1. INTRODUCTION

Wireless sensor networks are such a scientific tool by which we can understand the elements of the real world even from far off places. Initially it was required by armed forces but later they are used by civil applications because of their diverse nature. It is such a vast and diverse field that it is impossible for a single discipline to cover the whole field. Application experts, software engineers and control engineers altogether sum this field up completely. The application of wireless sensor networks (WSNs) is very rapidly expanding to our daily life in recent years. Since the basic operation of WSNs is data fusion, time synchronization is required for basic communication among nodes in WSNs. To fulfill this need all the nodes are brought to agree on a common clock. It does not necessarily mean that all the nodes act at once since this would become a very strict form of synchronization. Any node at any time can match its relative time with another node. Energy efficiency, robustness, cost effectiveness; life time and accuracy are all prerequisites of time synchronization. We propose that cognition is also a factor which should be required for time synchronization. It is not an easy task to get maximum benefits from sensor nodes because of certain factors like limited energy and smaller hardware structure. The timing information of sensors is mismatched because of the drifting of internal clocks of the sensors. As a result of this situation achieving of accurate information becomes difficult and thus useful transmission cannot be had. Good time synchronization is required to achieve maximum benefits.

WSNs cannot function over a long period of time because of energy deficiency and environmental factors. Moreover, during synchronization and messages exchange between two sensor nodes some quantity of energy is also consumed. As the sensor nodes remain unattended for quite some time and their batteries cannot be changed. In order to increase the life of a sensor network, it is mandatory to reduce its energy consumption or to develop fault tolerant, simple and accurate time synchronization protocols so that to make network functional in case of node failure due to less energy at the end sensor node. Overhearing is the type of synchronization which helps to decrease the dissipation of energy but at the same time it compromise on accuracy. Receiver only synchronization scheme is described in section two.

Accurate and cognitive protocols are required to solve this issue. WSNs should remain functional even in the presence of faults, only then we can get benefits from them. We call this state fault tolerance. And if the protocol removes the faults by itself and adjusts the network we call it cognition. Cognition, in a way, is self-healing process. In the present day time there does not exist any technique which can overcome all types of faults. We need accurate and low complexity protocols to preserve network functionality over a longer period of time.

Since the basic purpose of the WSNs operation is data fusion, therefore to fulfill this need all the nodes are brought to agree on a common time because this is the basic requirement of the any application of WSNs. Moreover, to agree local clocks of all sensor nodes exactly on a common time is a strict form of clock synchronization which is almost impossible to achieve. Due to the environmental factors and age of the local oscillator, local clocks of the sensor nodes may slightly drift from each other that's why they are measured as parts per million. So the rate of oscillations of two sensor nodes cannot be exactly same. Similarly due to fixed and random communication delays nodes confront difference of time also. So most often the local clocks are precisely matched. In this article we presented precise clock synchronization taking in account that local frequencies of sensor nodes are slightly different from each other and value of computed offset shared between the nodes also suffers little delay.

Cognitive global clock synchronization algorithm (CGCSA) adopts centralized synchronization mechanism. Because it is easy to control centralized synchronization schemes to achieve maximum benefits but the major drawback in centralized approaches is the failure of the master node. If we have a global time synchronization agreement between the sensor nodes which can handle faults also it provides us accurate and long term clock synchronization. As self-organization is a fundamental characteristic of the WSNs so we can direct this basic characteristic of WSNs for the cognition/fault tolerance.

The rest of the paper is organized as: Some important factors directly related to the clock synchronization and the CGCSA are discussed in section 2, previous related research work is represented in section 3, section 4 describes the description of the cognitive global clock synchronization (CGCS), measurement model of the CGCS is discussed in section 5, section 6 discusses simulation results, and finally research work is concluded in the section 7.

2. PLENARIES OF WSNS AND CLOCK SYNCHRONIZATION RELATED WITH CGCSA

WSNs fall in the category of distributed systems as many processors working together fulfill a certain application in the network communication. But WSNs are a little different from other distributed system because their communication range is comparatively shorter and their hardware structure is smaller. As WSNs are different from other distributed system with reference to their hard ware infrastructure and application, so they are faced by different problems from those of the distributed systems. In this section we will discuss those constraints which are a challenge in designing of the protocols in WSNs. If these challenges are handled and addressed properly, we can get a great help in increasing the performance efficiency not only in the existing protocols but also in new protocol designing.

2.1. Constraints in WSNs

Batteries are the energy source for the WSNs. If the battery of a node is discharged completely then this node does not remain functional. When nodes are operational a considerable amount of energy is consumed in data processing and communication. In the applications where energy efficiency is primary goal and accuracy can be compromised to some extent, they need resource efficient and energy efficient protocols. When one or more than one nodes stop working in a network this creates a bottleneck in fetching the data. To overcome this difficulty we need to design fault tolerant/cognitive protocols.

When sensor nodes are temporarily influenced by certain elements the communication is affected to some extent. There are some elements which completely disturb the functioning of sensor networks, these elements are temperature, humidity and chemicals. Actually these constraints affect the quality of the communication link. To solve this problem we need robust protocols.

Network density is also an important factor, effective coverage range can be short but area of coverage can be vast for which there is dense deployment of sensors and when the deployment is dense, it causes delaying factor because of multipath. Therefore effective routing protocols are required to solve this problem. Routing has two popular types (i) active routing and, (ii) proactive routing.

2.2. Coverage and Connectivity

Numbers of nodes in WSNs depend upon two factors: cost of WSNs and functionality. Coverage and connectivity are also important justifications of WSNs. Both of these come into existence because of limited communication. The deployment of sensors in different applications is of different nature. The most recent research has created the common impression that triangular form of deployment covers more geographical area than rectangular form of deployment of sensor nodes. It is because overlapping area is smaller in rectangular lattice and it needs fewer sensors. "Coverage" is such a quality of service which tells us for how long a period and in how much better way can the sensor networks be functional in the sensing field. When sensor nodes have covered any physical field then it is required to transform the sensed data into useful information by fusing that data, and to achieve it clock synchronization in sensor nodes is a necessary factor. When this data is transmitted to the sink node it is called "connectivity". To achieve useful and better coverage sensors are neither deployed much densely nor too far at long distances. The deployment of sensors can be random as well as fixed.

2.3. Network Topology

Beside with the coverage and connectivity in the sensor network, special attention is given to the network topology also. Network topology may be star, tree, single hop or multi hop. Coverage may be sparse, dense or redundant. Connectivity may be connected or sporadic. Topology control is an extremely important issue without it nodes cannot function in any way for a longer period of time. It is quite difficult to better the performance of network without affecting its capacity and connectivity. If hop by hop short range method is applied instead of long range communication method, the possibility of the packet data loss is reduced to a considerable extent. For synchronization in mobile sensors dynamic network topology is used to make clock synchronization more robust, while in stationary sensors they are tightly coupled according to the desired application.

2.4. Clock Synchronization

Clock synchronization is an operation in which it is ensured that all physical distributed processors are functioning at a common notion of time. This function is useful for security and fault tolerance. The data of all nodes is collected in order to arrange a meaningful result. That is why clocks are synchronized at a common time. There are several types of clock synchronization; however we will discuss only those schemes which are used in or directly related with our cognitive global clock synchronization scheme.

Master-slave synchronization: In this type there is a master node which is equipped with CPU processor and all the children nodes have to set their local clocks at the local time of the master node.

Peer-peer synchronization: In this scheme every node can communication independently with any other node. In this type the possibility of the failure of master node is lessened but to control this scheme is considerably difficult. Self-organization is a basic quality of sensor networks. Through this quality it is easy to control master-slave technique and getting maximum benefits. Therefore considering these factors we have overcome the biggest danger of failure of master node resulting the communication being affected, we overcome this problem by putting cognition (inclusion of expected master node) in it.

Sender-receiver synchronization (S-R): One node sends the time stamp to another node and the other node records it sends back the acknowledgement to the first node and the message delay time is measured by calculating round trip time. Communication between master node and expected master node in CGCSA is followed by this strategy. Traditional S-R synchronization strategy adopted by [5] is shown in Figure 1.

Receiver-receiver synchronization: In this approach when two receivers receive the same timing message, they mutually compute the difference of their reception time sent by their common sender. The major advantage of this scheme is the reduction of message delay variance. Communication between expected master node and the children nodes is followed by this strategy. The R-R synchronization scheme exploited by [4] is represented in Figure 2.

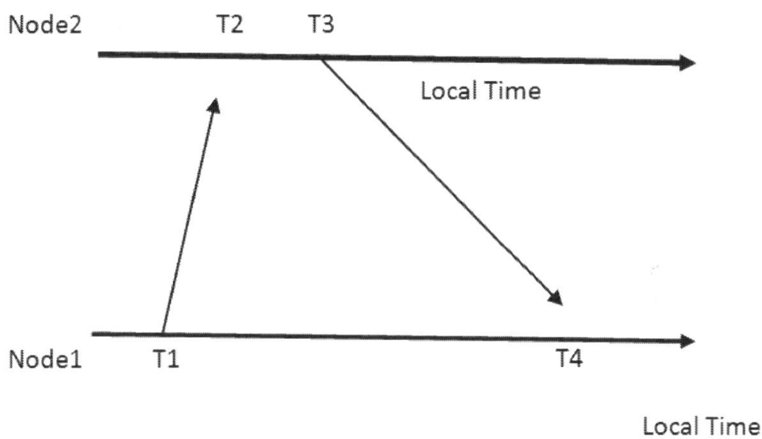

Figure 1. Sender-receiver synchronization

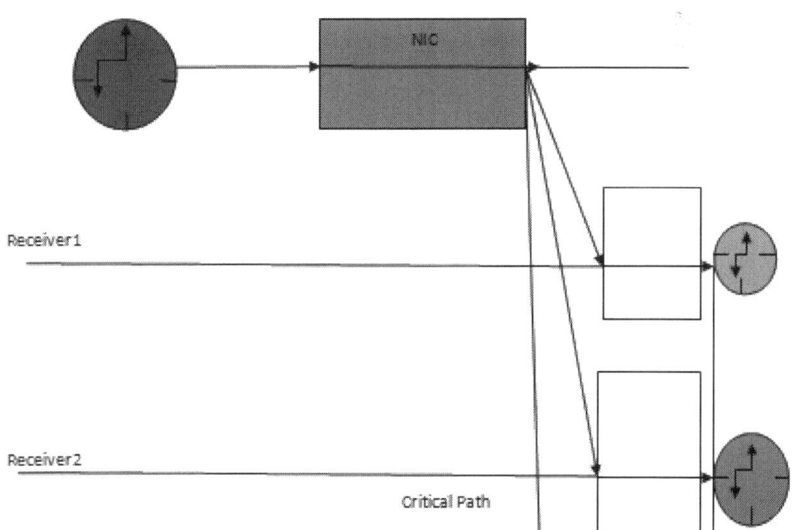

Figure 2. Traditional receiver-receiver synchronization

Receiver only synchronization: In this approach two super nodes communicate with each other for the synchronization and the third node residing nearby overhears all the messages and synchronizes itself upon reference time without communicating directly. As it can be seen from Figure 3 that node M and node N exploit pairwise synchronization and node C just overhears the message packets exchange between the nodes and synchronizes itself with the root node. Communication overhead is very low in this scheme and it saves a lot of energy but this scheme does not work efficiently for the networks dealing with large number of sensor nodes and secondly direct communication is always more accurate than the overhearing case. Another disadvantage of this scheme is that all the nodes are not active all the time. If node(s) is in idle state then it will miss the synchronization packet to overhear.

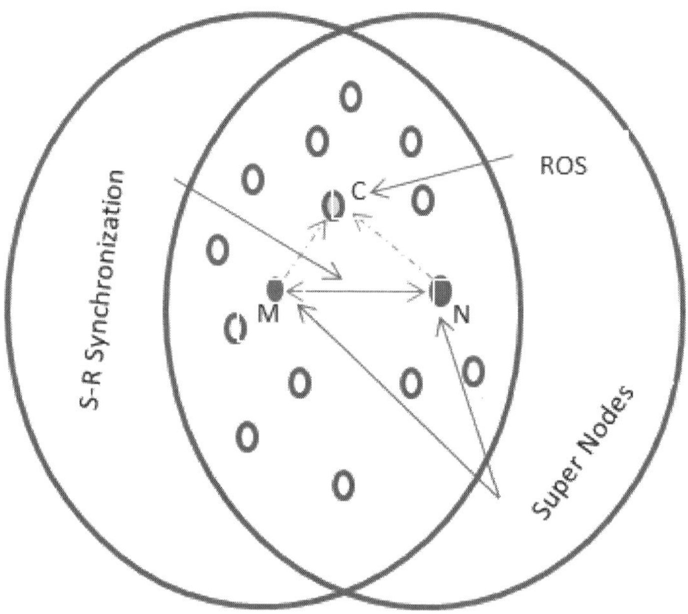

Figure 3. Receiver only synchronization

To understand clock synchronization properly it is necessary to understand in detail the clock skew side by side with the clock offset. As the sensors are equipped with the very cheap clocks and if their internal clocks drift among themselves the synchronization period cannot sustain for a long period of time. All the processing devices are equipped with digital clocks and this hardware clock causes the time synchronization to hold or disturb. Each hardware clock has two components: quartz stabilizer oscillator and the counter. Counter is incremented upon rising or falling edge. Rate of the counter is defined as the derivative of the counter with respect to real time. The rate of an ideal clock would be unity at all-time but a number of environmental factors like temperature, humidity, big age of crystal etc. cause the drifting of clocks among themselves. That is why clocks are measured in parts per million.

3. LITERATURE REVIEW

The word sensor means to perceive. Sensor is a device which converts a physical phenomenon into a signal to enable an observer to read it. Sensors are used to detect the electrical and optical signals and these act by converting the physical parameters into signals which in turn may be read electrically. Sensors are also known as a mote. A mote is always a node; however, a node is not always a mote. This is a tiny device with a small hardware and very limited memory. It runs on a minute battery. The sensor nodes gather and process the information before communicating it to the network. Sensor nodes being smaller and cheaper, and because of their various usages are rapidly making their place in the present day scientific technology like MEMS (Micro electro mechanical system). We can monitor the desired condition at any location with the help of these small sized devices. Sensors follow multi-hop communication when they are deployed at different locations.

A node's transmission range varies depending upon the protocol used in the transmission. Due to the limited resources, WSNs faces various challenges such as self-management, power restriction, storage restrictions, security and random topology. Multiple sensor nodes combine to make a network. Some of these nodes may act as control nodes. These nodes are wirelessly connected with each other and with the base station. A node is either a sensor node or a router node in nature. Base station is a platform where all the nodes send any form of information. Sensor nodes are very densely deployed. These follow multi hop communication and function like networks. Owing to the vulnerability, their security and reliability is a big challenge. Energy consideration, scalability, data aggregation and fault tolerance are the key features of WSNs protocol designing.

A sensor network is comprised of sensor nodes, sink, sensor field and controller. Sensor nodes are the most important component of this network. They act by gathering and routing the information before transferring it to the sink. The information is transferred in very few messages leading to reduced energy consumption.

WSNs are comprised of various sensor nodes which are densely deployed, few of them are control nodes also referred as base stations. These nodes are connected with each other wirelessly and base station connects them with some other network. Sensors are deployed in ad-hoc fashion to aggregate data from observing field. Less number of sensor nodes is deployed in an ad-hoc network than a sensor network. Nodes deployment may be random or deterministic, for deterministic deployment routs are predefined while for premier case it's really very challenging. A sensor network is comprised of sensor nodes, sink, sensor field and controller. Sensor nodes are the most important component of this network. They act by gathering and routing the information before transferring it to the sink. The information is transferred in very few messages leading to the accuracy. WSNs have both advantages as well as disadvantages. Among the advantage, their implementation cost is less, the network may be deployed anywhere and they can be monitored through a global monitoring system. The disadvantages include inadequate security, difficulty in configuration and reduced speed as compared to a wired network.

To aggregate data in a sensor network, clock synchronization is an important feature. Synchronization has been studied for long time for computer systems but traditional schemes are ineffective for sensor because of their different physical properties. Because of less energy at end sensor node it's hard to synchronize nodes on common clock effectively and due to power consumption factor running the complex synchronization algorithms is difficult. In wired networks, network time protocol and global positioning system are useful methods for clock synchronization but are not suitable for wireless network because of their different characteristics. There are many factors that must be taken into account while designing a clock synchronization algorithm in sensor networks like energy efficiency, scalability, precision, robustness, life time, scope, and cost and size, thus a single protocol may not satisfy altogether. Clock synchronization is useful for communication between nodes in the network. It is required for proximity, location and also important due to imperfections in local oscillators due to which an event is not recorded at the same time.

Wireless sensor networks have variety of applications ranging from domestic [14-16] to scientific fields [17-19], and purpose of communication is data processing [9, 10]. Looking into the needs it's required to synchronize all nodes to a global clock because the basic operation is data fusion. Sensor network applications require that sensors in the network agree on the common time. By matching the sensor location and the sensing time, the sensor network may easily predict the transmitted information. But this is not an easy task as clock skew and clock offset disturb clock synchronization a lot. Node's hardware oscillator shows clock as approximation $Ci(t)$ of real time t as

$$C_i(t) = \omega_i(t) + \phi_i \tag{1}$$

Where ω_i denotes the rate of hardware oscillator and ϕ_i denotes the time difference. Due to the imperfections and environmental changes rate of clock can be expressed as

$$1-\rho \leq \omega \leq 1+\rho \tag{2}$$

Sensor, often called motes are valuable tiny devices used to interact with physical world and to achieve useful information from a long distance accurately. They are often deployed in ad hoc fashion. Basic purpose is data fusion and data from each sensor is collected for a meaningful output [11-13]. Data fusion not only defines sensors location but proximity with each other also, for efficient data fusion clock synchronization is the basic requirement. Due to various limitations and bottle necks clock synchronization is hardly achieved. So far many protocols have been introduced in this field like RBS [4], TSPN [5], FTSP [6], Tiny-Sync [7], and Firefly Synchronization in Ad Hoc Networks [8]. Almost all protocols follow more or less the same timing messages exchange mechanism as followed by the pioneer protocols of the field [4] and [5], these protocols use receiver-receiver (R-R) and sender-receiver (S-R) synchronization schemes for clock synchronization respectively and are quite simple to implement.

Sensor nodes have limited resources and are prone to faults [1]. Sensor hardware and communication should be fault tolerant so that network remains functional [21]. The life time of the sensor network is increased by activation of sensors on the basis of residual energy content in [2]. A distributed algorithm is presented in [7] to overcome the faults. In [26] Li et al solved joint estimation problem by maximizing profile likelihood function in terms of clock skew. As sensors are prone to faults so fault tolerant clock synchronization protocols in sensor networks are required to be fault tolerant, so that if some node goes faulty it may be discarded or replaced by some other node.

4. THE CGCSA DESCRIPTION

Our proposed scheme works in two phases i.e. Synchronization and cognition.

4.1. Synchronization and Cognition

Since sensor nodes are deployed in far flung areas therefore fault tolerant and accurate transmission is needed. Considering the needs we have designed CGCS protocol so that we can produce a simple, accurate and efficient time synchronization protocol. In our design the node with high energy level will be selected as the master node and if any child node fails to function our algorithm will discard it, as shown in Figure 4. And if master node or expected master node fails to function, this matter will be dealt with logically and intelligently. In other words, if the expected master node fails the master node will check its table and selects a new expected master node, and incase the master node fails, the expected master node will start functioning as the master node and will select its expected master node. The master node will form its table with the help of information provided to it by acknowledgement (ACK) messages from all the nodes.

Nodes with different residual energy may present in the sensor network. Nodes with low energy cannot contest for the master node in CGCSA. Communication between master and expected master node ensures the energy state of main controlling node(s). The likelihood is a better tool for checking the performance of fault tolerance protocols. The master node and the expected master node exchange a periodic message which helps to keep the alternative paths alive in order to maintain the network reliability in case of any faults. Excessive amount of energy and bandwidth is consumed during the master node's connection with the rest of the nodes. Therefore, like the unicast protocol, in our protocol the master node will only keep itself linked with the expected master node.

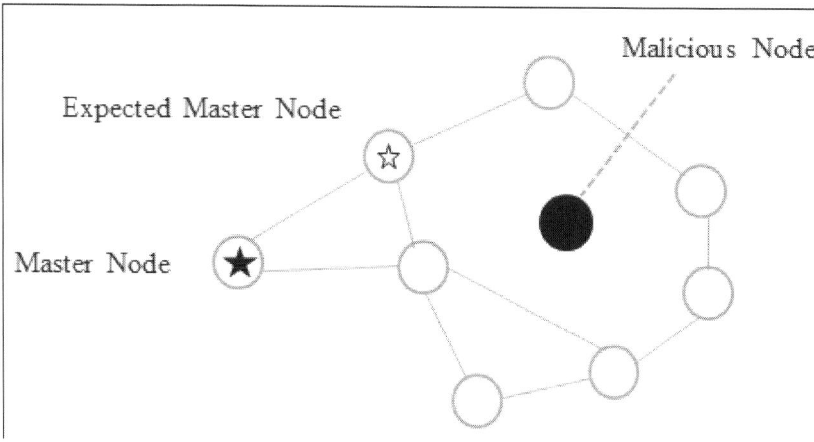

Figure 4. Synchronization and cognition

It can be seen from Figure 5 that the node which receives timing information form master node compares the clock value for adjustment, computes the offset and sends it to next hop. This mechanism reduces message overhead and in turn propagation time is also reduced. It is not essential for any individual receiver to send back the ACK to its sender except to the master node. Meanwhile master node and expected master node communicate with each other to check their effectiveness. If node "N" is unable to receive message from master node then node "N" appears as master node and broadcasts the request message to the next hop to change and update their receiver ID. On the other hand if master node "M" is unable to receive ACK message from "N" continuously three times then it means that node "N" is down.

So the parent node checks its table and sends a request message to "C" to communicate with it as a secondary expected master node. Each node sends an ACK message to the parent node, due to centralized system round trip time is minimized. ACK massage includes ID information and energy level of the sensor nodes. Only parent node maintains the table which contains the ID information and clock time of all the nodes. CGCSA adopts S-R synchronization and R-R synchronization schemes so achieves benefits of both strategies at the same time i.e. accuracy and energy efficiency at some extent.

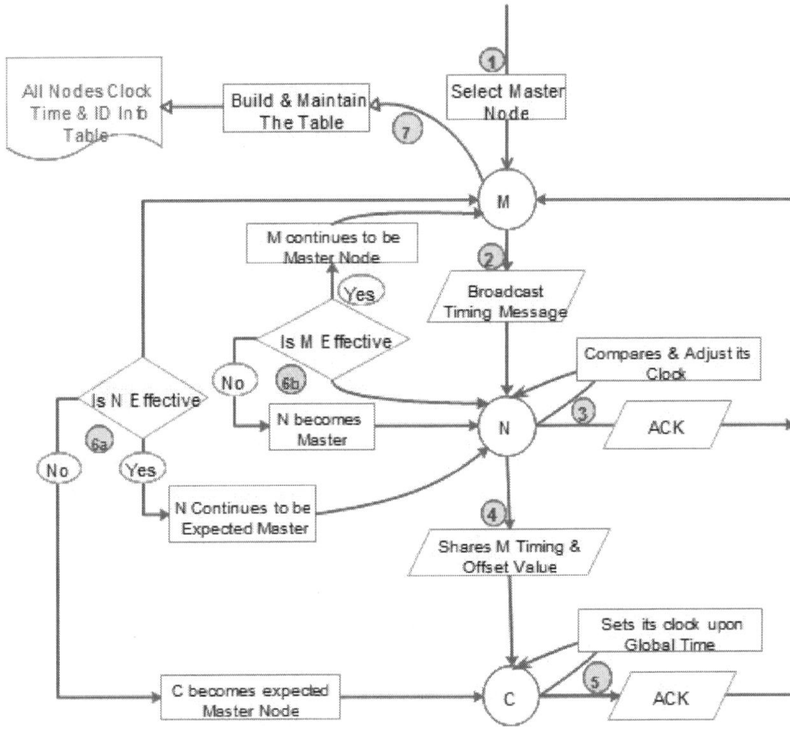

Figure 5. Block diagram of the algorithm

5. CLOCK MODEL FOR THE CGCS

The CGCS algorithm proposed is a new approach in clock synchronization domain to deal with faults in sensor networks and to achieve accurate clock synchronization for meaningful data fusion. The algorithm not only synchronizes the sensor nodes on mutually common time but also checks for the faulty nodes and takes decisions likewise.

Due to environmental factors these tiny sensor nodes cannot survive malicious attacks in hostile environments [20, 22]. Implementation of cognitive algorithm is essential to discard malicious nodes and if master node gets faulty another node must take its place immediately so that the network remain synchronized. Taking an account on all bottlenecks like limited energy, smaller bandwidth and presence of faulty nodes, our aim turned into a novel clock synchronization algorithm which not only checks for the faults but also encounters with those faults very well.

Beside of sender to receiver (S-R) synchronization and receiver to receiver (R-R) synchronization there exists another approach which is called as receiver only synchronization (ROS) [25]. In ROS a node overhears synchronization messages of two super nodes without communicating by itself, due to reduced transmissions this scheme saves energy.

Message exchange mechanism for the CGCSA is illustrated in Figure 6.

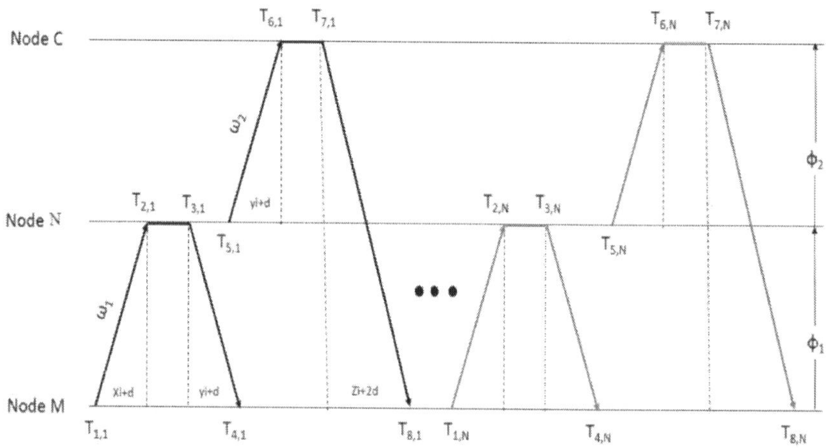

Figure 6. Clock synchronization model of the CGCSA

CGCSA assumes S-R and R-R approaches that exploit the timing exchange mechanism, as depicted in Fig. 3. Send time and receive time estimate round trip time (RTT) and RTT is exploited to synchronize the clocks in the network.

Figure 6 presents a number of N message exchanges between the nodes. The timestamps can be expressed as

$$T_{2,i} = \omega_1 (T_{1,i} + x_i + d) + \phi_1 \tag{3}$$

$$T_{3,i} = \omega_1 (T_{4,i} - x_i - d) + \phi_1 \tag{4}$$

$$T_{6,i} = \omega_2 (T_{5,i} + y_i + d) + \phi_2 \tag{5}$$

$T_{7,i} = \omega_2(T_{8,i} - z_i - 2d) + \phi_1 + \phi_2$

(6)Where $T_{1,i},\dots\dots,T_{8,i}$ are the time stamps between the three nodes, ϕ_1 is the clock offset between node M and node N while ϕ_2 is the clock offset between node N and the node C. Similarly ω_1 is the clock skews present between node M and node N and ω_2 is the clock skew existing between node N and node C. d is the fixed propagation delay and x_i, y_i, z_i are the random delays. Latency and jitter affect the performance of clock synchronization to a great extent. Fixed delays can be compensated easily but the latency changes with the passage of time. Random delays which can cause end to end latency are divided into four categories: send time, access time, propagation time and receive time.

Equations (3), (4), (5) and (6) can be written in the form of random delays, like

$$x_i = \frac{T_{2,i} - \phi_1}{\omega_1} - T_{1,i} - d \tag{7}$$

$$x_i = -\frac{T_{3,i} - \phi_1}{\omega_1} + T_{4,i} - d \tag{8}$$

$$y_i = \frac{T_{6,i} - \phi_2}{\omega_2} - T_{5,i} - d \tag{9}$$

$$z_i = -\frac{T_7 - 2\phi_1}{\omega_2} + T_8 - 2d \tag{10}$$

When a sensor node sends a message to another node it confronts mainly these types of delays: (a) send time, (b) access time, (c) propagation time, and (d) receive time. Sensor nodes synchronize with one another by sharing their present state. Clocks can agree on mutually common time through global positioning system (GPS) but this is an expensive solution. If GPS is used then there is no use of WSNs being cheap. Moreover, if sensors are deployed inside a building then clocks cannot be synchronized on GPS because of nonexistence of line of sight. The protocol introduced by us not only achieves synchronization but also deals with faults. Clock synchronization becomes a difficult task because of less energy at the end sensor node; as the life of a node depends on its battery life. Sensors are disposable when their energy ends they cease to function. So fault tolerant clock synchronization protocols are mandatory to achieve accurate data fusion for long time.

The random delays from equations (7), (8), (9) and (10) are subject to exponential distribution. The joint maximum likelihood function for the above equations can be written as

$$L(d,\omega_1,\omega_2,\phi_1,\phi_2) = \lambda^{-4N} \exp\left[-\frac{1}{\lambda}\sum_{i=1}^{N}[2x_i + y_i + z_i\right] =$$

$$\left[\lambda^{-4N} \exp -\frac{1}{\lambda}\sum_{i=1}^{N}\left[\left(\frac{T_{2,i}-\phi_1}{\omega_1} - T_{1,i} - d\right) + \left(-\frac{T_{3,i}-\phi_1}{\omega_1} + T_{4,i} - d\right) + \left(\frac{T_{6,i}-\phi_2}{\omega_2} - T_{5,i} - d\right) + \left(-\frac{T_{7,i}-2\phi_1}{\omega_2} + T_{8,i} - 2d\right)\right]\right] \quad (11)$$

Taking derivative of the equation (11) w.r.t $\omega_1,\omega_2,\phi_1,\phi_2$ and for getting the value of each individual parameter all other parameters are fixed and result is set equal to the 0. After derivation we get

$$\hat{\phi}_1 = \frac{\sum_{i=1}^{N}\left[\omega_1'\left(T_{2,i}+T_{3,i}\right)-\left(T_{1,i}+T_{4,i}\right)\right]}{2N\hat{\omega}_1'} \quad (12)$$

$$\hat{\phi}_2 = \frac{\sum_{i=1}^{N}\left[\left(T_{5,i}+T_{8,i}-2d\right)\times\left(T_{6,i}{}^2+T_{7,i}{}^2\right)-\left(T_{6,i}+T_{7,i}\right)\times Q\right]}{\sum_{i=1}^{N}\left[\left(T_{6,i}+T_{7,i}\right)\times\left(T_{5,i}+T_{8,i}-d\right)\right]-2NQ} \quad (13)$$

Where

$$Q \cong \sum_{i=1}^{N}\left[\left(T_{1,i}T_{2,i}+T_{3,i}T_{4,i}\right)+\left(T_{2,i}-T_{3,i}\right)d\right]$$

$$\hat{\omega}_1=\frac{\sum_{i=1}^{N}\left(T_{2,i}+T_{3,i}\right)-2N\hat{\phi}_1}{\sum_{i=1}^{N}\left(T_{3,i}+T_{4,i}\right)} \quad (14)$$

$$\hat{\omega}_2=\frac{\sum_{i=1}^{N}\left(T_{6,i}+T_{7,i}\right)-2N\hat{\phi}_2}{\sum_{i=1}^{N}\left(T_{5,i}+T_{8,i}-d\right)} \quad (15)$$

MSE is a better performance index of an estimator. We purpose estimator which outperform ML estimators in terms of MSE. Numerical simulations are carried out for the clock model. The results are represented and discussed briefly in the next section.

6. SIMULATIONS AND DISCUSSIONS

In this section we discuss the numerical simulations which are carried out through MATLAB depending upon the measurement model to justify the performance of our proposed clock synchronization scheme. There exist three famous types of clock synchronization strategies i.e. S-R synchronization followed by TPSN [5], R-R synchronization adopted by RBS [4] and receiver only synchronization (ROS) presented in [25]. As CGCSA achieves synchronization as the hybrid of S-R and R_R synchronization so its performance is compared with the third type which is ROS scheme.

The parameters set up for the simulations are represented in Table 1.

Figure 7 and Figure 8 illustrate the mean square error (MSE) of the estimated clock offset ϕ_1 and ϕ_2 versus the number of rounds of messages exchange. Curves in both plots for the values (1, 1.10, and 1.80) for ω_1 and ω_2 clearly show that the MSE of the estimated clock offset show decreasing pattern and convergence as long as the number of rounds of messages exchange increase and it establishes the effectiveness of proposed estimator. It can also be seen in Figure 7 and Figure 8 that MSE of the clock offset is directly proportional to the clock skew values i.e. lesser is the clock skew value , lower will be the MSE of the estimated clock offset and vice versa. MSE of the clock offset is lowest without clock skew error.

Table 1. Synchronization parameters

Parameter	Symbol	Values
Clock offset	ϕ_1 and ϕ_2	1.5-2
Clock skew	ω_1 and ω_2	1-1.80
Fixed delay	d	0.1-0.8
Number of observations	N	30-35

The MSE of the estimated clock skew ω_1 and ω_2 versus the number of rounds of messages exchange is shown in Figure 9 and Figure 10 respectively. The fixed propagation delay d is set to 0.1ms, 0.4ms and 0.8ms. We can see from both the figures that MSE of the clock skew converges with the passage of timing rounds. Distance between the nodes is directly concerned with the fixed propagation delay. Higher is the fixed propagation delay, longer is the distance and higher is the MSE of the estimated clock skew.

Performance of our proposed scheme is compared with receiver only synchronization (ROS) and results are presented in Figure 11 and Figure 12 in terms of clock offset and clock skew estimation for the equal set of values. The basic reason of comparing with ROS scheme was that in ROS strategy 3rd node residing nearby overhears without communication itself, while in CGCSA the 3rd node actively takes part in communication by receiving and sending the messages. Although it increases message overhead as compared to the overhearing and consumes more energy than ROS but active participation of all nodes increases the accuracy. Simulations results show the better performance of CGCSA. From simulation results we can conclude that:

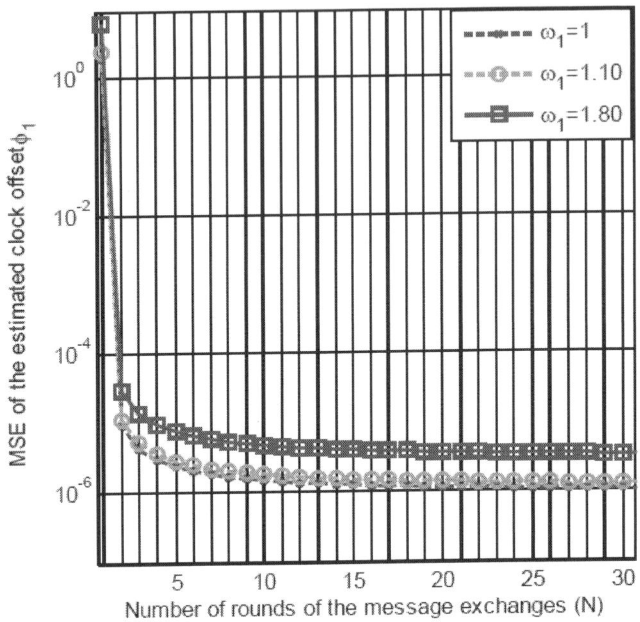

Figure 7. MSE of the estimated clock offset between node M and node N versus number of rounds of message exchanges

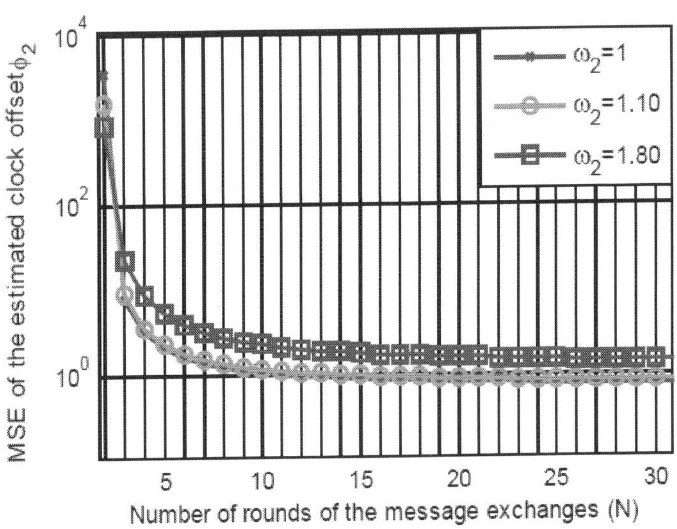

Figure 8. MSE of the estimated clock offset between node N and node C versus number of rounds of message exchanges

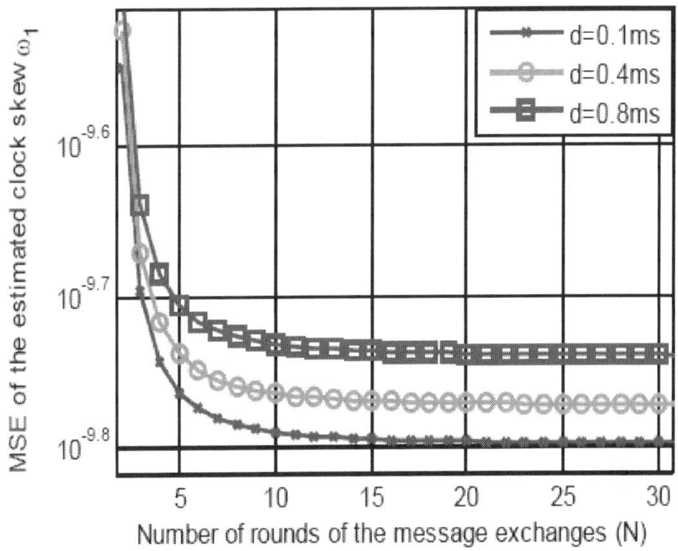

Figure 9. MSE of the estimated clock skew ω_1 versus number of rounds of message exchanges

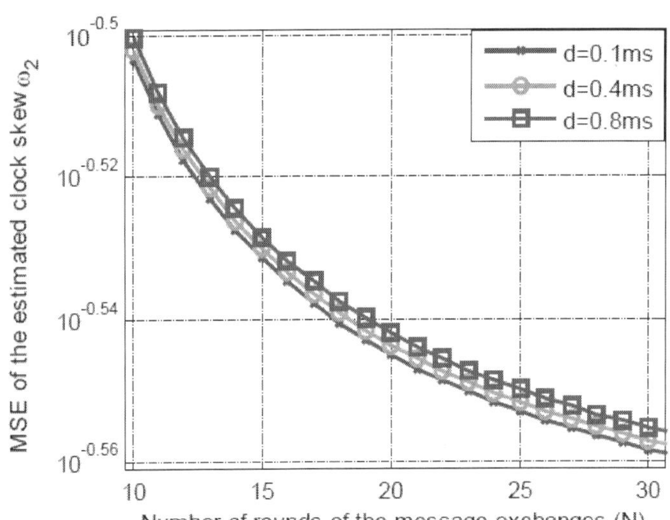

Figure 10. MSE of the estimated clock skew ω_2 versus number of rounds of message exchanges

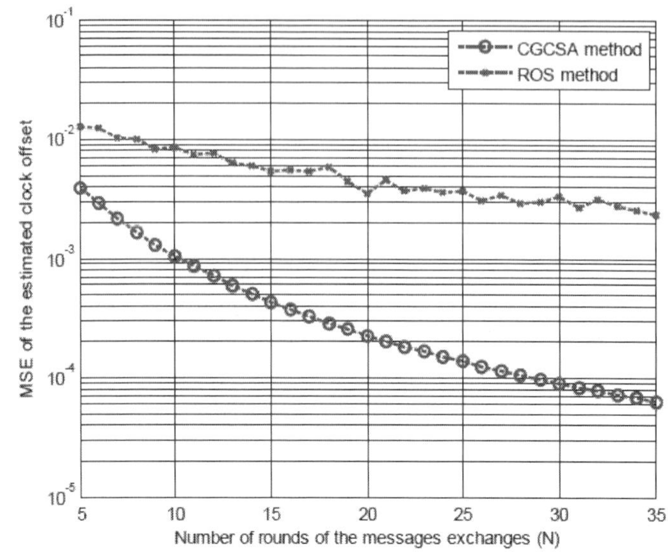

Figure 11. Comparison of the MSEs of the clock offset of proposed scheme with ROS scheme

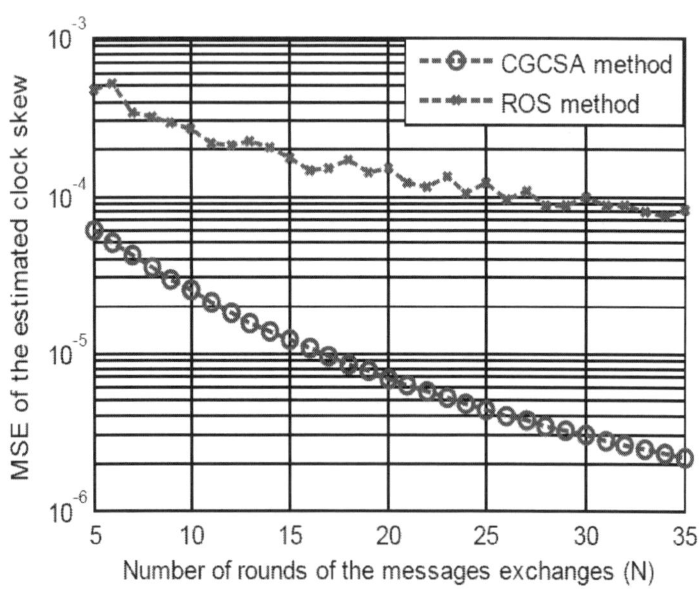

Figure 12. Comparison of the MSEs of the clock skews of proposed scheme with ROS scheme

Convergence of lower bounds equal to zero reflects three aspects: (a) asymptotic efficiency, (b) lesser number of transmissions required for synchronization via CGCS and (c) accuracy.

No doubt ROS scheme is more energy efficient due to the overhearing of the third node but it compromises accuracy but our scenario is accurate because all nodes actively take part in synchronization.

This paper presented a new clock synchronization scheme in WSNs whose main idea is to share estimated value as a reference for the network wide synchronization to achieve global agreement, this reduces burden on the master node and saves on energy of the master node by not forwarding the requests from master node. Its novel fault tolerant characteristics make it robust to dynamic network topologies. From the results and the novelty of the CGCSA we have found some characteristics which are summarized in Table 2.

Table 2. Basic characteristics of CGCSA

Deployment	Manual
Mobility	Immobile
Energy	Battery
Modality	Radio
Infrastructure	Ad-hoc
Topology	Single hop, multi hop, graph
Quality of service	Real time
Fault tolerance	Novelty of cognition
Scalability	Achievable through clustering
Accuracy	High
Application	Flexible for any wherever accuracy is desired e.g. for target tracking and law enforcement applications

Our algorithm not only detects faults in message exchanges but also replaces the faulty nodes and makes logical decisions as shown in Fig. 2. The algorithm rely on bilateral synchronization between master node and expected master node possibly with a tree topology and it follows unilateral synchronization between master node and the child node(s) based on a reference clock value shared the expected master node.

Lemma 1: Let N_{CGCS} be the required timing messages in CGCS then $N_{CGCS} = 3N + L(L-1)/2$, where L stands for the number of overall sensors in the network and N stands for the number of messages exchange between the nodes.

Proof: Since every node in the network is connected to the master node and there are $L-1$ branches in hierarchical tree. $2N$ Timing messages are required for synchronization between the master node and the expected master node, moreover expected master node shares time readings upon message received from master node to the child node to compensate relative clock offset between them. Thus $N_{CGCS} = 3N + L(L-1)/2$.

7. CONCLUSIONS

In this article useful and important feature cognition has been introduced and we study joint estimation of clock skew and offset in RTS mechanism. We investigated JMLE estimator. This paper combines the performance of RTS and a brief introduction of our fault tolerant novel algorithm. The proposed scheme is fully distributed, includes skew compensation even for the higher values of the clock skew like 1.80, computationally simple, accurate and easy to implement. It is obvious from the simulations that it reduces end to end latency. We have compared our algorithm with ROS scheme and found it more accurate. Message complexity is low in CGCSA and as a result accuracy is high. Clustering phenomenon is required for its implementation in larger networks. Same algorithm can be used for inter clustering synchronization and intra clustering synchronization, it will reduce energy consumption.

As our future work we are focusing on energy consumption during synchronization periods and designing of the CGCSA with re-election the Ex-master node as a Master node. There are various interesting open research directions for future work. First the assumption of fixed d is same for all links is imprecise for the real sensor network. Second estimation of communication link delays will be quite interesting.

ACKNOWLEDGMENTS

This work was financially supported by Shanghai Science and Technology Foundation (13510500400).

REFERENCES

1. J. Lyer, H. Yu, H. Kim, E. J. Kim, K. Hawan, P. S. Mah, Assuming k-coverage in the presence of mobility and wear-out faults in wireless sensor networks, IJSNet., vol .5, no. 1, pp. 58-65, 2009.

2. S. He, J. Chen, D. K. Yau, H. Shao, Y. Sun, Energy efficient capture of stochastic events under periodic network coverage and coordinated sleep, IEEE Trans. Parall. Distrib. Syst., vol. 23, no. 6, pp. 1090-1102, 2012.

3. C. Caione, D. Brunelli, L. Benini. Distributed compressive sampling for lifetime optimization in dense wireless sensor networks, IEEE Trans. Ind. Inform., vol. 8, no. 1, pp. 30-40, 2012.

4. J. Elson, L. Girod, and D. Estrin, Fine-grained network time synchronization using reference broadcasts, ACM SIGOPS Operating Systems Review, vol. 36, no. SI, pp. 147-163, 2002.

5. S. Ganeriwal, R. Kumar, M. B. sirivastava, Timing-sync protocol for sensor networks. In Proceedings of the 1st international conference on Embedded networked sensor systems, ACM, pp. 138-149, 2003.

6. M. Maróti, B. Kusy, G. Simon, A. Ledeczi, The flooding time synchronization protocol, In Proceedings of the 2nd international

conference on embedded networked sensor systems, ACM, pp. 39-49, 2004.

7. S. Yoon, C. Veerarittiphan and M. L. Sichitiu, Tiny-sync: Tight time synchronization for wireless sensor networks, ACM Transactions on Sensor Networks, vol. 3, no. 2, pp. 1-34, 2007.

8. Tyrrell, G. Auer, C. Bettstetter, Fireflies as role models for synchronization in ad hoc networks. In proceedings of the 1st international conference on Bio inspired models of network, information and computing systems, ACM, pp. 1-7, 2006.

9. J. Hill, M. Horton, R. Kling, The platforms enabling wireless sensor networks, Communications of the ACM, vol. 47, no. 6, pp. 41-46, 2004.

10. Autonomous sensing and communication in a cubic millimeter, http://robotics. eecs. berkeley. edu/pister, Smart Dust, 2008.

11. Dobra, M. Garofalakis, J. Gehrke, R. Rastogi, Processing complex aggregate queries over data streams, In proceedings of the ACM SIGMOD international conference on Management of data, pp. 61-72, 2002.

12. S. Madden, R. Szewczyk, M. J. Franklin, and D. Culler, Supporting aggregate queries over ad-hoc wireless sensor networks Mobile Computing Systems and Applications, In proceedings Fourth IEEE Workshop on IEEE, pp. 1-9, 2002.

13. Woo, S. Madden, R. Govindan, Networking support for query processing in sensor networks. Communications of the ACM, vol. 47, no. 6, pp. 47-52, 2004.

14. R. P. Council. Maryland Department of Transportation, Managing the Transportation System, 1983.

15. P. Johnson, D. Andrews, Remote continuous physiological monitoring in the home. Journal of telemedicine and telecare, vol. 2 no. 2, pp.107-113, 1996.

16. Q. Li, M. D. Rosa, and D. Rus, Distributed algorithms for guiding navigation across a sensor network. In proceedings of the 9th annual international conference on Mobile computing and networking, ACM, pp. 313-325, 2003.

17. K. Sohraby, D. Minoli, T. Znati, Wireless sensor networks: technology, protocols, and applications, John Wiley & Sons, 2007.

18. V. Ashok, S. Sutagundar, S. Manvi, Context aware multisensor image fusion for military sensor networks using multi-agent system, vol. 2, no. 1, pp. 147-167, 2011.

19. K. Shanmugavalli, and K. Fathima, Wireless Sensor Network Based Surveillance System to Track Enemy Intrusion at Borders, vol. 3, no. 1, pp. 138-144, 2014.

20. K. Sun, et al., Secure and Resilient Time Synchronization in Wireless Sensor Networks. Springer, pp. 347-367, 2007.

21. Zhang, Stabilizing clock synchronization for wireless sensor networks, University of Iowa, pp. 335-349, 2007.

22. J. H. Hoepman, A. Larsson, E. M. Schiller, and P. Tsigas, Secure and self-stabilizing clock synchronization in sensor networks. Stabilization Safety and Security of Distributed Systems Springer, pp. 340-356, 2007.

23. L. Villas, G. L. Daniel, G. Maia, W.P. Richard, Jó Ueyama, A.F.L. Antonio, An energy efficient joint localization and synchronization solution for wireless sensor networks using unmanned aerial vehicle, Wireless Networks, pp. 1-14, 2014.

24. J. Lee, J. Kim, E. Serpedin, Clock offset estimation in wireless sensor networks using bootstrap bias correction, Wireless Algorithms Systems and Applications, Springer, pp. 322-329, 2008.

25. K-L Noh, E. Serpedin, and K. Qaraqe, A new approach for time synchronization in wireless sensor networks pairwise broadcast synchronization, IEEE Transactions On wireless Communication, vol. 7, no. 9, pp. 3318-3322.

26. J. Li and D. R. Jeske, Maximum likelihood estimation of clock offset and skew under exponential delays, Appl. Stoch. Models Bus. Ind., 25, pp. 445-459, 2009.

7

PERFORMANCE OF HMM BASED BURST ERROR MODELING FOR CDMA AND OFDM TRANSMISSIONS

Rakesh Ranjan[1], Debjani Mitra[2]

[1]Department of Electronics and Communication Engineering, National Institute of Technology (NIT), Patna, 800005, India

[2]Department of Electronics Engineering, Indian School of Mines (ISM), Dhanbad, 826004, India

ABSTRACT

The availability of efficient, fast and accurate burst error models are useful for evaluating and improving higher communication protocols in many of the current generation networks using the basic code division multiple access (CDMA) and orthogonal frequency division multiplexing (OFDM) blocks. We estimate in this paper discrete HMM error models for these two important transmissions in frequency selective fading. Statistics like average loglikelihood, average length of error burst, autocorrelation function, distribution probability of error free intervals and average error probabilities have been applied to judge the accuracy of the models and estimate their optimum order and parameters. The order of a HMM is defined as the minimum number of hidden states to model the error statistics accurately. The performance of the modeling algorithm has been analysed and evaluated with respect to the input sample period of the signal on which the channel acts, and also on crucial system parameters like number of interferers in CDMA and subcarrier size in OFDM.

KEYWORDS

CDMA, OFDM, Hidden Markov Model, Baum Welch Algorithm, Discrete Channel Modeling

1. INTRODUCTION

Wireless broadband capabilities and services are offering increased data transfer rates and lower latencies in the current generation. In an effort to render ubiquitous wireless connectivity a reality, the contribution of several technology advances is significant. Multiple Input/Multiple Output (MIMO), Space Time Coding (STC) using higher order of modulation and encoding schemes, beam forming and beam directionality control, interference mitigation and others are some to name. Multicarrier communication realized by OFDM systems achieving frequency diversity is quite adequately established for broadband communications. DS-CDMA has emerged as one of the most prominent access scheme for the air interface of 3G cellular systems from capacity and network planning considerations. Convergence of CDMA-OFDM is the latest paradigm in the evolution and design of wireless networks [1-4].

Accurate models of network dynamics are needed for analysing and evaluating the performance of existing network protocols in the wireless environment without implementing the complex physical layer and long conditional probability calculations. The design and performance evaluation of interleaving and FEC schemes in particular are cumbersome if approached through analog channel models and waveform level simulations. Discrete or Finite State Channel Modeling, (DCM) on the other hand models the temporal correlations and the statistical properties of bursty errors encountered in wireless channels with memory. Simple state diagrams are capable of characterizing the input-output characteristic of the channel and can accurately regenerate a bursty error sequence not only in terms of the average error probability, but also in terms of size and frequency of the error bursts. These fast and efficient channel models reduce the cost and time of simulation of the whole physical channel and hence the availability of accurate DCMs can be a powerful tool for rapid experimentation and prototyping.

Initiated by Gilbert, his proposed model determined the capacity of wireline channel with burst-error [5], and the basic idea evolved with several extensions and modifications like the Gilbert-Elliott channel (GEC) model, Fritchman model, McCullough Models and few others [6-8]. These models have several recent applications in protocol analysis in varied applications [9, 10]. Since the last thirty years, the statistical framework of Hidden Markov Models (HMM), very lucidly described by Rabiner [11] is successfully being used in several applications requiring pattern recognition. Naturally, HMMs have also been proposed for the accurate representations of the statistical patterns typical of bursty error data in fading channels [12-14].

Performance of HMM based error modeling validated against measurement values or offline waveform level simulations have been presented in the literature for some specified standard wireless systems, mostly with the purpose of testing of higher layer protocols [15-19]. The choice of the number of states is highly dependent upon the physical conditions of the wireless channel, length of error sequence, modulation schemes used, input conditions of the system, etc. The order of a HMM is defined as the minimum number of hidden states required to model the error data accurately. Results related to HMM order estimation as has been addressed in [20, 21] is in a generalized approach and not for any specific system. For OFDM systems, the HMM order estimation has been currently addressed in [22]. The authors have derived the order of hidden Markov discrete channel models for 4 different input sample periods. In [23], a radio channel emulator based on HMM is presented and validated for a WCDMA system for reproducing the lower layers of the UMTS radio interface, the Physical Layer, the Data Link Layer and some functionality of the Network Layer. They have computed the probability distribution function of the number of errors in a packet, length of burst error as well as error free intervals for their validation.

Verikoukis [24] has obtained adaptive HMM for indoor OFDM based physical layers and HMMs have been modelled even upto 16 states to match the length of error free intervals in the offline simulated burst error sequences. In their work on estimating the channel state in a CDMA system employing AMC and multicodes,[25] it has been shown that the HMM filter is robust and provides a significant throughput improvement over the Simple Moving Average filter, especially when the channel estimate is quite noisy or the normalized Doppler rate is small in Rayleigh fading. The results presented in [26] related to frame level error models for GSM channels clearly depict that different models have different performance depending on the order of the statistical metric being compared. Some approach may be efficient for matching the correlation structure of the bursty periods, but poor in reproducing the autocorrelation or reproducing the error rate. For some characteristics, semi hidden Markov modelling is more appropriate while for satisfying the others, even 4th order HMMs with 5 or 6 states have been experimented upon.

The literature study clearly indicates that performance evaluation of HMM modelling cannot be generalized at all, as it is strongly dependent on the system being emulated, the fading channel and the type of discrete channel modelling approach adopted. To the best of our knowledge, the combined consideration of the channel as well as the emulation of basic transmission links for the estimation and performance evaluation of HMM modelling has not received much attention in the open literature. Against this motivation, the current work has investigated the performance of HMM based DCM for two popular emerging technologies for broadband wireless systems, namely CDMA and OFDM, which are the most commonly employed building blocks in advanced wireless access networks and international wireless standards. The basic CDMA/OFDM block needs a DCM representation in several low level applications for which the availability of a simple first order HMM with lower number of states is often more required and the objective of the current work is to derive such simple models that can provide reasonably good accuracy.

The rest of the paper is organized as follows. Section 2 discusses the general HMM implementation trained by the Baum Welch algorithm as applied for the emulated CDMA and OFDM links. The system simulation parameters for these two systems have been specified here for which the models have been developed. The results regarding HMM performances and order (required minimum number of HMM states) estimations for these two transmission schemes are presented in the Section 3 and finally conclusions are summarized in Section 4.

2. HMM IMPLEMENTATION FOR DCMS

HMMs are powerful statistical modeling tool, extensively implemented for the modeling and analysis of error burst in discrete communication channels. For a wireless discrete channel, a hidden Markov model with N states and K symbols, for a discrete channel is described by the triplets:

$$\Gamma = \{A, B, \Pi\} \tag{1}$$

where, A is an $N \times N$ state transition matrix that gives the state transition probabilities between N states, B is an $K \times N$ error probability generation matrix while being in a particular state, and Π is an $1 \times N$ matrix, called state probability matrix, and it gives the probability of being in a particular state at the start of the process. These parameters can be estimated using an iterative procedure, known as the Baum-Welch Algorithm (BWA). The implementation of BWA involves the assumption of initial model, $\Gamma_i = \{A_i, B_i, \Pi_i\}$ and the computation of two different probabilities, the forward path probability (α) and the backward path probability (β) using the Forward-Backward algorithm[27]. The Baum-Welch algorithm is designed to converge to the maximum likelihood estimator of model $\Gamma = \{A, B, \Pi\}$ that maximizes $Pr(0^{*}|1)$, where, $\overline{O}$ is the error sequence obtained through simulation, with $O_k = 1$ indicating that the transmission error for the kth transmitted bit, and $O_k = 0$ representing the error-free transmission. The number of iterations, for convergence, can be obtained by continuing the simulation for several iterations, until the variation in the value of $Pr(\overline{O}|\Gamma)$ become very small. For larger data size, forward and backward vectors may exponentially become zero. Therefore, to prevent numerical underflow, (α) and (β) vectors must be scaled properly. The scaling constant C_t, is defined as,

$$C_t = \sum_{i=1}^{N} \alpha_t(i) \tag{2}$$

where, $t = 1, 2, \ldots, T$ bits and $i = 1, 2, \ldots, N$ states.

For large error sequences, this probability will be very small and hence it is conventional to use the log-likelihood ratio L, defined as:

$$L = \log_{10} Pr[\overline{O} \mid \Gamma] = \sum_{t=1}^{T} \log_{10} C_t \tag{3}$$

where, $t = 1, 2, \ldots, T$ bits.

2.1. CDMA Transmission Links

CDMA is an interference-limited spread spectrum multiple access technique. As all the users transmit on the same frequency, internal interference generated by the system is the most significant factor in determining system capacity and call quality. The HMM based model for CDMA links as simulated in the current work is shown in Figure 1(a). The spreaded symbols from the desired user and the interfering users are transmitted to the physical channel. The channel simulation includes the effects of Multiple Access Interference (MAI), multipath fading and Additive White Gaussian Noise (AWGN). After despread and demodulation, the received data are compared with the source data, which generates the error sequence due to imperfections and impairments in the system. The obtained error sequence is then used to execute the HMM using the BWA training and a new error sequence is generated. The original and the HMM-generated error sequences are then compared in terms of several statistical measures.

2.2. OFDM Transmission Links

The OFDM model for hidden Markov discrete channel modeling has been shown in the Figure 1(b). It includes several blocks, such as, Modulation/Demodulation, Serial-to-Parallel/Parallel-to-Serial, Zero-Pad/ Remove zero-pad, Add/Remove cyclic prefix, IFFT/FFT, Physical channel. The guard interval at the start of each symbol decreases the overall spectral efficiency of the system but at the same time it eliminates the effect of ISI on an OFDM signal. The effect of both AWGN and Multipath-fading has been incorporated in the wireless channel. On comparing the data source and the received signal, the error sequence is obtained and this is due to the imperfections in transmitter, receiver, and channel. The error sequence is then trained by the Baum-Welch algorithm, which results in a new error sequence. The HMM generated error sequence and the original error sequences are compared.

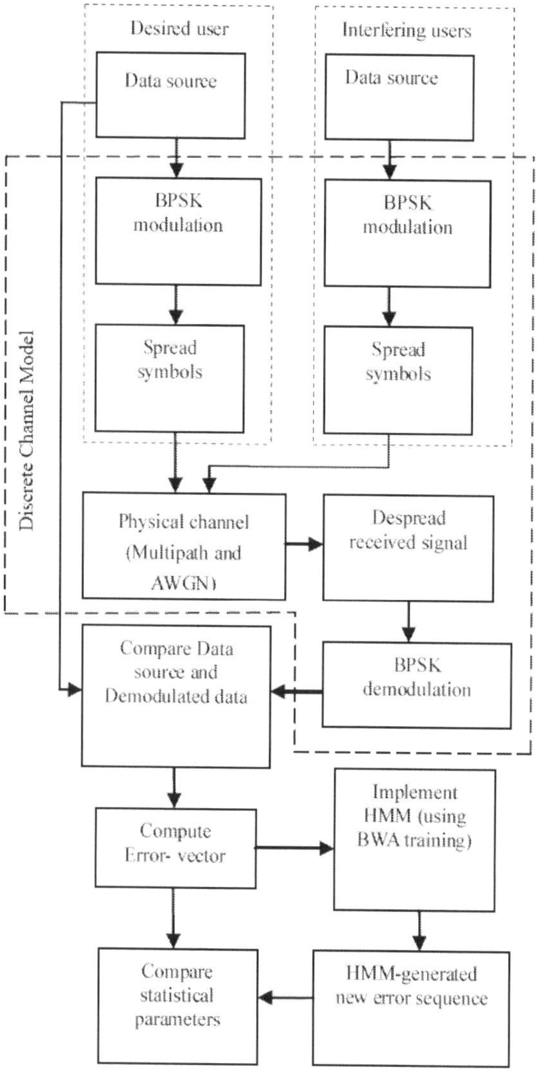

Figure 1(a). HMM implementation for DCM of CDMA links

3. SIMULATION RESULTS

HMM DCM for frequency selective multipath fading channel for CDMA and OFDM transmission systems have been derived here using MATLAB with the simulation parameters as listed in Table 1. The values of path delays and path gains are kept constant, as the variations in these values has very little effect on the number of bits in error for a particular value of input sample periods. The higher SNR value, assures that the generated error bursts are not due to AWGN only, but the multipath fading has the major contributions.

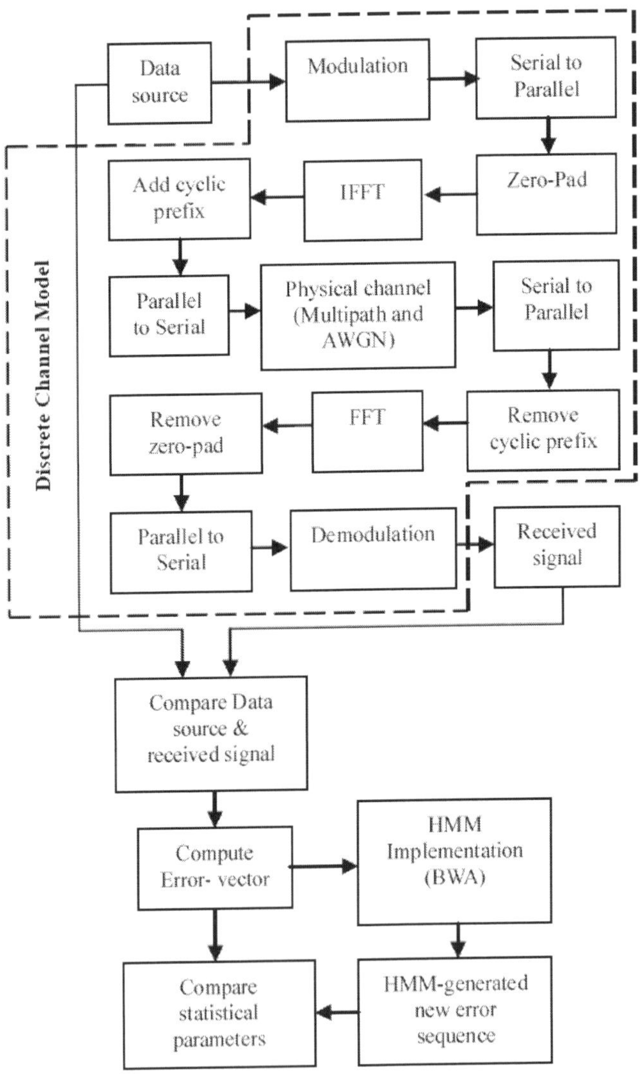

Figure 1(b). HMM implementation of DCM for OFDM links

Table 1. Simulation parameters

Maximum Doppler Shift (Hz)	60
Sample Period (T_S) of input signal on which the channel acts (μsec)	1000, 100, 10, and 1
SNR (dB)	0-20
Path-Delays of the discrete paths (μsec)	0, 1.5, 3.2
Average Path-Gains (dB)	0, -3, -5
Length of error sequence for BWA training	Approx. 100000
Number of iterations for BWA simulation	50
Modulation scheme for OFDM	8-QAM
Sub-carrier size of OFDM	52
Corresponding FFT size of OFDM	64
Modulation scheme for CDMA	BPSK
Spreading Factor for CDMA	15

3.1. HMM Model for CDMA System

For the CDMA system, the variation of average loglikelihood with the number of HMM-states under the varying conditions of input sample period (T_S) of the signal is shown in Figure 2. The physical conditions of the wireless channel can be realized in terms of input sample period (Ts) of the signal. The input sample period (Ts) of the signal is the sample period of the input signal on which the channel acts and it plays a vital role in wireless DCM. The low average loglikelihood value indicates that a lesser number of error bits have been generated. The figure indicates the superiority of HMM channel modelling for higher sampling rates of input signals, i.e. lower T_S values. Table 2 shows the average size of error bursts or error probability for different T_S values in the original and HMM-generated sequences. With lower Ts values, the fading channel is effectively acting for smaller intervals of time, leading to lower average error probability. The average length of error bits being smaller, the HMM training is capable of achieving higher loglikelihood of matching a given observed error sequence.

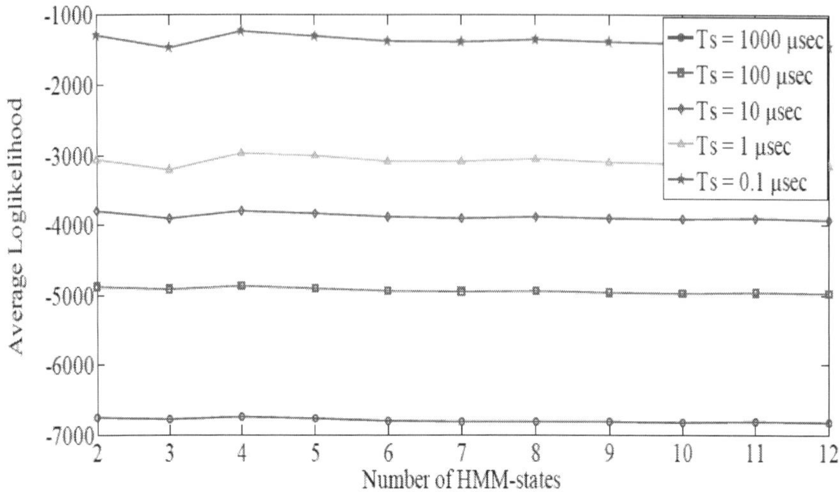

Figure 2. Average Loglikelihood comparison for CDMA system for different T_S values

The discrete wireless channel modeling approach makes use of N numbers of HMM states. For estimating the required minimum number of hidden states for accurate channel modeling, we compare the autocorrelation function and $\Pr(0^m \mid 1)$, for NI = 11, as shown in Figure 3 and Figure 4. The notation $\Pr(0^m \mid 1)$ symbolize the probability of the event of observing m or more consecutive error-free transmissions following an error, i.e. the probability of distribution of error-free intervals. Autocorrelation function (ACF) comparison is one of the most popular techniques for performance evaluation of HMM for finding the closeness between the original and HMM-generated data. From Figure 3(a), Figure 3(b), and Figure 3(c) a comparison of the ACF performance can be made with respect to the different values of T_S. From extensive simulation runs it was found that for T_S = 1000 µsec, 100 µsec, and 10 µsec, 2-state HMM was sufficient to model the wireless channel of CDMA system and there was no need to go for higher state HMMs. For T_S = 1 µsec, and 0.1 µsec, the ACF comparison shows that the orders of the Markov model are 6 and 8 respectively. The error-free interval distribution have been compared in Figure 4(a) and Figure 4(b), for T_S = 1 µsec, and 0.1 µsec, also emphasize that the required minimum number of HMM-states are 6 and 8 respectively.

It was also found that the required minimum number of HMM states for CDMA also rely upon the number of interfering users (NI). The required minimum HMM states for CDMA system under different channel and system conditions are summarized in Table 3. Table 4 shows the effect of numbers of training iterations of Baum-Welch algorithm on the order estimation, for SF = 15 and NI = 11. It shows that for higher T_S values, upto 10 μsec, number of iterations has no impact on order of HMM, while for lower T_S values (1 μsec and 0.1 μsec), on decreasing the number of iterations, usually it was observed that no suitable order can be obtained, as upto 12-state HMMs were simulated.

3.2. HMM Model for OFDM System

A similar simulative analysis for deriving the hidden Markov discrete channel models for OFDM transmission links is presented in this subsection. Here the impact of the input signal sample period is observed to be affecting the training performance in a similar manner as the CDMA system, as shown in Figure 5. The size of error bursts of the OFDM system, for different T_S values is tabulated in Table 5. Similar to CDMA, it was observed that, for the lower T_S values, the size of HMM generated error bursts were varying widely from the original sequence and by decreasing the T_S values, size of error bursts or error-probabilities decreases.

To estimate the minimum number of HMM states required for a given OFDM DCM model, comparison of ACFs and $\Pr(1^m|0)$, was applied. The notation $\Pr(1^m|0)$, characterize the probability of the event of observing m or more consecutive errors following an error-free transmission, i.e., the distribution of error intervals. Numerical simulations, for the OFDM system with 52-subcarrier was used to estimate the minimum number of hidden state, shows that 2-state HMM was sufficient to model the system quite accurately for T_S = 1000 μsec. For T_S = 100 μsec, 10 μsec and 1 μsec , the ACF comparisons, as shown in Figure 6(a), Figure 6(b), and Figure 6(c), illustrate that, the required minimum number of hidden states are 3, 6, and 8 respectively. The error-interval distribution comparison for T_S = 10 μsec and 1 μsec are shown in Figure 7 and the closeness of 6-state and 8-state HMM with their original error sequences, again accentuate that the required minimum number of hidden state is 6 and 8 respectively.

Table 2. Comparison of size of error bursts of CDMA for different T_S values in the original sequence and HMM-generated sequences

NI	Error Sequence	T_S = 1000 µsec	T_S = 100 µsec	T_S = 10 µsec	T_S = 1 µsec	T_S = 0.1 µsec
11	Original	3598	2398	2124	1921	578
	2-state	3581	2332	1995	1858	484
	5-state	3588	2441	2244	1402	170
	8-state	3683	2386	1901	1720	751

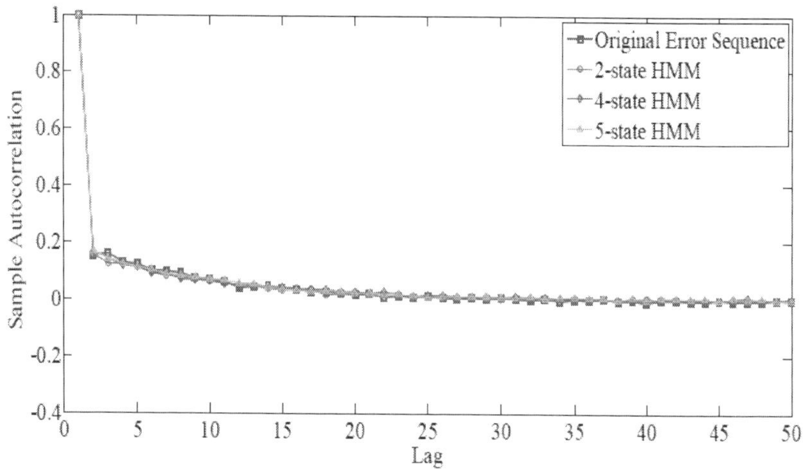

Figure 3(a). Comparison of autocorrelation functions of 2-state, 4-state and 5-state HMMs with original error sequence for the input sample period of 10 μsec for CDMA

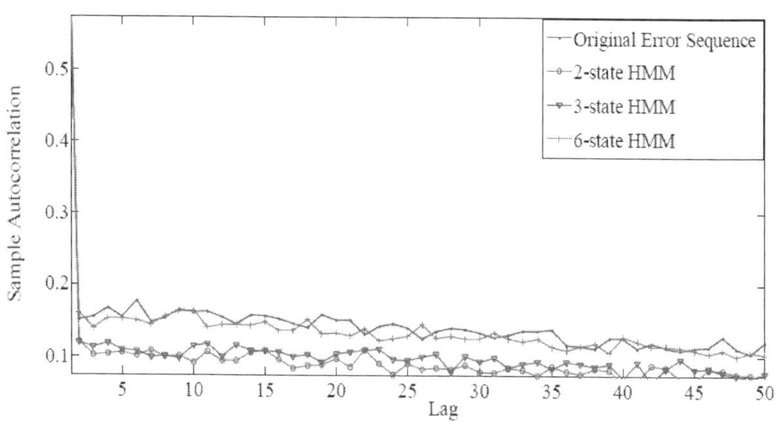

Figure 3(b). Comparison of autocorrelation functions of 2-state, 3-state, and 6-state HMMs with original error sequence for the input sample period of 1μsec for CDMA

The orders of HMMs for OFDM systems under different channel and system conditions are tabulated in Table 6. Finally, the estimated parameters $[A]$,$[B]$, and $[\Pi]$ of HMMs for both the wireless systems are given in Table 7 and Table 8 for different T_s values. As these estimated parameters are strongly characterized by the error bursts produced in the wireless fading channel, therefore it can be effectively used in design and analysis of several applications and protocols.

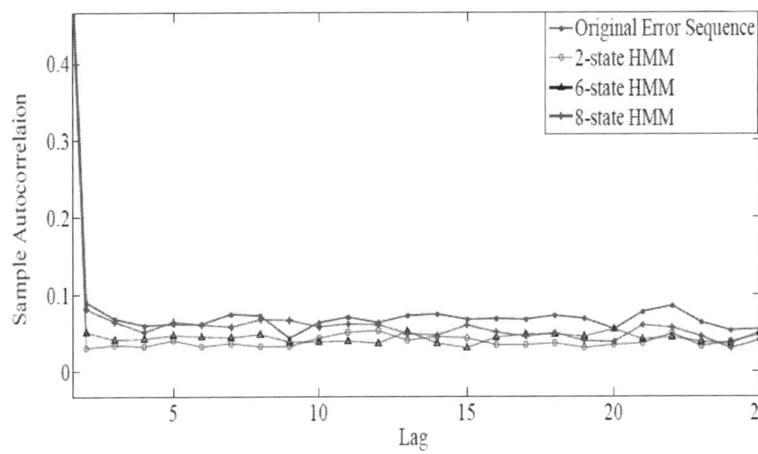

Figure 3(c). Comparison of autocorrelation functions of 2-state, 6-state, and 8-state HMMs with original error sequence for the input sample period of 0.1 μsec for CDMA

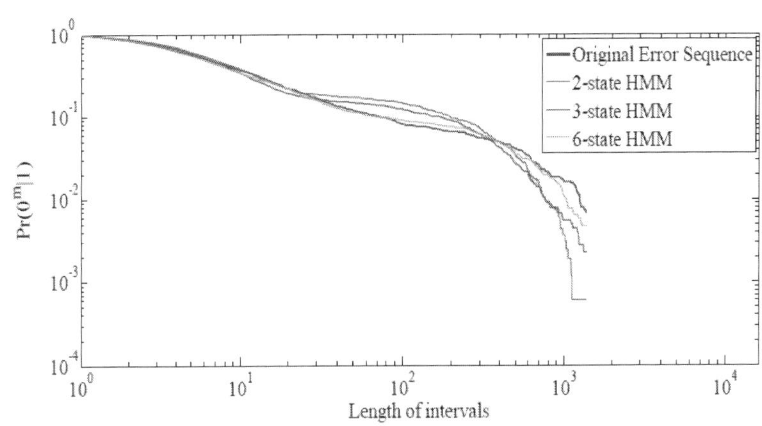

Figure 4(a). Comparison of error-free interval distribution of original and HMM-generated error sequences for Ts = 1 μsec for CDMA

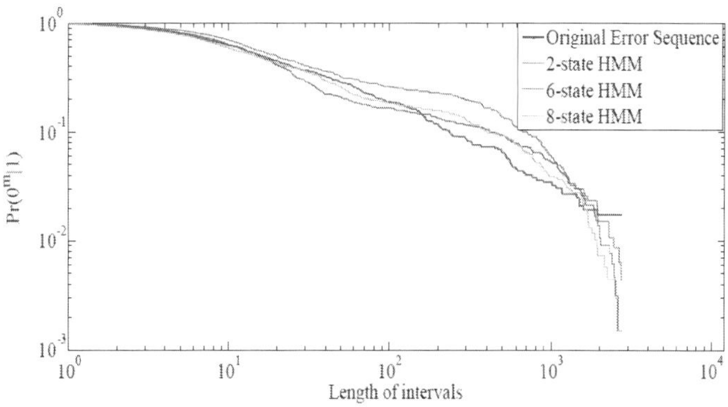

Figure 4(b). Comparison of error-free interval distribution of original and HMM-generated error sequences for Ts = 0.1 μsec for CDMA

Table 3. Summary of required minimum number of hidden states for CDMA (SF = 15)

Number of interfering users (NI)	Input Sample Period (T_S) (μsec)	Required minimum number of HMM states
11	1000	2-states
11	100	2-states
11	10	2-states
11	1	6-states
11	0.1	8-states
04	1	4-state
04	0.1	5-state

Table 4. Effect of number of iterations for CDMA (SF = 15 and NI = 11) on th e HMM order

T_S (μsec)	Number of iterations				
	50	40	30	20	10
1000	2-state	2-state	2-state	2-state	2-state
100	2-state	2-state	2-state	2-state	2-state
0	2-state	2-state	2-state	2-state	4-state
1	6-state	12-state	#	#	#
0.1	8-state	#	#	#	#

- upto 12-state HMMs are simulated, but no close match found

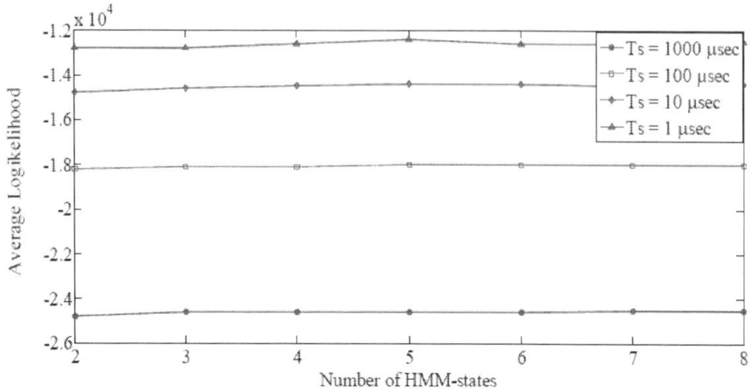

Figure 5. Average Loglikelihood comparison for OFDM system for different Ts values

Table 5. Comparison of size of error bursts of OFDM for different Ts values in the original sequence and HMM-generated sequences

Sub - carrier size	Error Sequence	T_S = 1000 μsec	T_S = 100 μsec	T_S = 10 μsec	T_S = 1 μsec
52	Original	25152	16286	12559	9629
	2-state	25201	16397	12781	9900
	5-state	24689	16081	13922	14014
	8-state	25124	16272	12785	8655

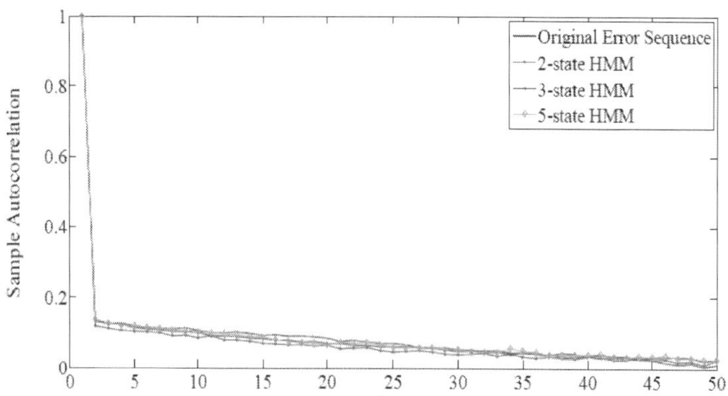

Figure 6(a). Comparison of autocorrelation functions of 2-state, 3-state and 5-state HMMs with original error sequence for the input sample period of 100 μsec for OFDM

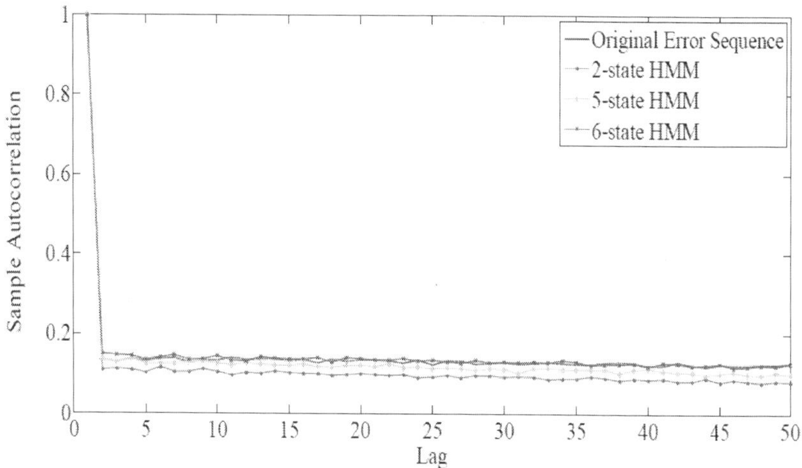

Figure 6(b). Comparison of autocorrelation functions of 2-state, 3-state and 6-state HMMs with original error sequence for the input sample period of 10 μsec for OFDM

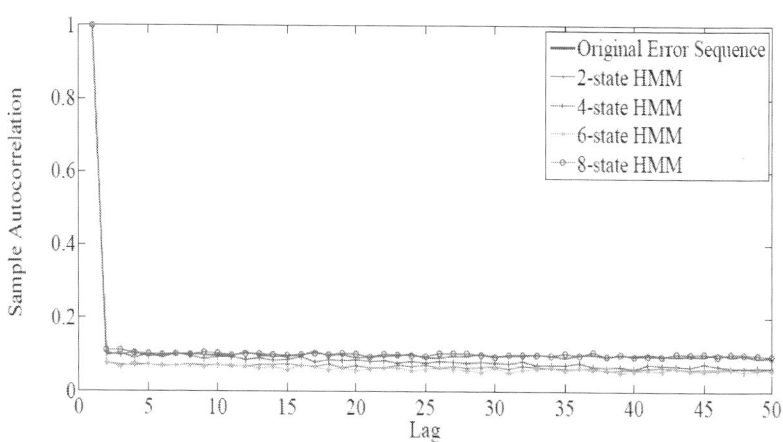

Figure 6(c). Comparison of autocorrelation functions of 2-state, 4-state, 6-state, and 8-state HMMs with original error sequence for the input sample period of 1 μsec for OFDM

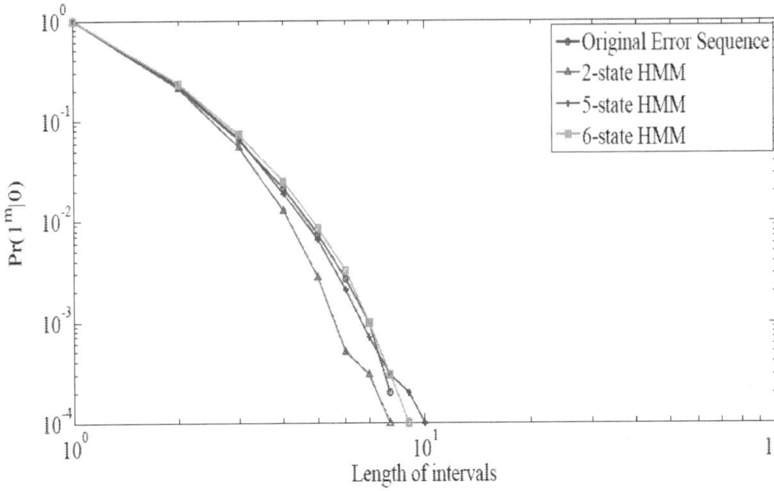

Figure 7(a). Comparison of error-interval distribution of original and HMM -generated error sequences for T_s = 10 μsec

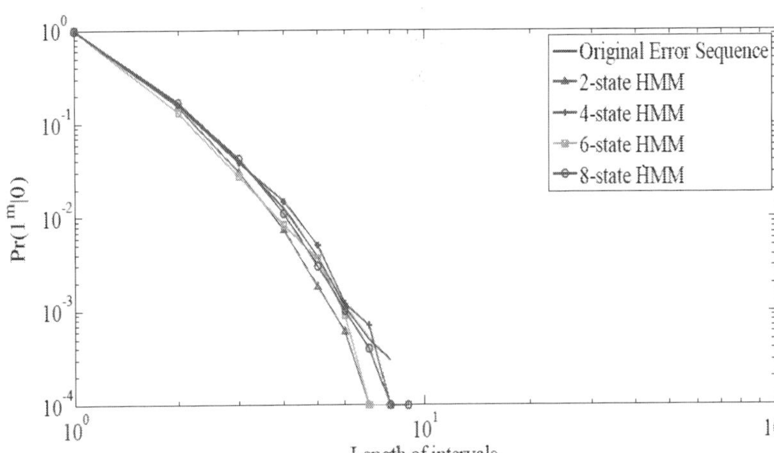

Figure 7(b). Comparison of error-interval distribution of original and HMM -generated error sequences for Ts = 1 μsec for OFDM

Table 6. Summary of required minimum number of hidden states of HMMs for OFDM

Sub-carrier size	FFT size	Input Sample Period T_S (μsec)	Required minimum no. of HMM states
52	64	1000	2-states
52	64	100	3-states
52	64	10	6-states
52	64	1	8-states

Table 7. Estimated parameters of HMMs for CDMA transceiver system, with SF = 15, NI = 11, and SNR = 10 dB

<table>
<tr><th>T_S (μsec)</th><th colspan="2">[A]</th><th colspan="4">[B]</th><th colspan="3">[Π]</th></tr>
<tr><td>100</td><td colspan="2">0.9298 0.0702
0.1539 0.8461</td><td colspan="4">0.9966 0.9309
0.0034 0.0691</td><td colspan="3">1.0000 0.0000</td></tr>
<tr><td>10</td><td colspan="2">0.9894 0.0106
0.0923 0.9077</td><td colspan="4">0.9984 0.8066
0.0016 0.1934</td><td colspan="3">1.0000 0.0000</td></tr>
<tr><td rowspan="10">1</td><td rowspan="7">[A]</td><td>0.8818</td><td>0.0866</td><td>0.0255</td><td>0.0058</td><td>0.0003</td><td>0.0000</td><td>0.0000</td></tr>
<tr><td>0.1338</td><td>0.7842</td><td>0.0763</td><td>0.0053</td><td>0.0004</td><td>0.0000</td><td>0.0000</td></tr>
<tr><td>0.0344</td><td>0.1221</td><td>0.8286</td><td>0.0138</td><td>0.0010</td><td>0.0000</td><td>0.0000</td></tr>
<tr><td>0.0122</td><td>0.0790</td><td>0.1588</td><td>0.7096</td><td>0.0393</td><td>0.0010</td><td>0.0000</td></tr>
<tr><td>0.0004</td><td>0.0011</td><td>0.0043</td><td>0.0212</td><td>0.8975</td><td>0.0755</td><td>0.0000</td></tr>
<tr><td>0.0000</td><td>0.0000</td><td>0.0000</td><td>0.0003</td><td>0.0431</td><td>0.9493</td><td>0.0072</td></tr>
<tr><td>0.0000</td><td>0.0000</td><td>0.0000</td><td>0.0000</td><td>0.0000</td><td>0.0162</td><td>0.9838</td></tr>
<tr><td rowspan="2">[B]</td><td>0.9994</td><td>0.9992</td><td>0.9986</td><td>0.9874</td><td>0.9645</td><td>0.9359</td><td>0.7264</td></tr>
<tr><td>0.0006</td><td>0.0008</td><td>0.0014</td><td>0.0126</td><td>0.0355</td><td>0.0641</td><td>0.2736</td></tr>
<tr><td>[Π]</td><td>0.5542</td><td>0.3386</td><td>0.1071</td><td>0.0000</td><td>0.0000</td><td>0.0000</td><td>0.0000</td></tr>
</table>

Table 8. Estimated parameters of HMMs for OFDM transceiver system, with sub-carrier size = 52, FFT size = 64, and SNR = 10 dB

T_S (μsec)	[A]	[B]	[Π]
1000	0.9723 0.0277 0.0280 0.9720	0.8650 0.6501 0.1350 0.3499	0.9943 0.0057
100	0.9775 0.0198 0.0027 0.0207 0.9633 0.0160 0.0036 0.0301 0.9663	0.9749 0.8376 0.6139 0.0251 0.1624 0.3861	0.0000 1.0000 0.0000

T_S (μsec)	Parameter	Values
10	[A]	0.9009 0.0894 0.0096 0.0001 0.0000 0.0000 0.2017 0.7656 0.0323 0.0004 0.0000 0.0000 0.0230 0.0844 0.8613 0.0313 0.0000 0.0000 0.0000 0.0001 0.0054 0.9911 0.0031 0.0002 0.0000 0.0000 0.0000 0.0120 0.8184 0.1696 0.0000 0.0000 0.0000 0.0006 0.1330 0.8664
	[B]	0.9945 0.9733 0.9273 0.8792 0.7868 0.5796 0.0055 0.0267 0.0727 0.1208 0.2132 0.4204
	[Π]	0.9947 0.0053 0.0000 0.0000 0.0000 0.0000

4. CONCLUSIONS

In this paper, HMM based discrete error models for frequency selective faded wireless channels has been estimated for two popular emerging broadband technologies namely CDMA and OFDM, which happen to be the building blocks in advanced wireless access networks and international wireless standards. Efficient estimation of the parameters of the HMM (including its number of states) has been approached by the powerful BWA iterative training and the statistics of the HMM generated error sequences have been extensively analysed and validated for both the transmission schemes with respect to several statistical measures. The input sample period (T_S) of the signal on which the channel acts, is seen to play a significant role in the performance evaluation of the HMM for both these systems and the corresponding minimum number of states required have been obtained in each case. The effect of decreasing the input signal sample period reduces the minimum number of hidden Markov states required to accurately model and analyse the error bursts in both CDMA as well as OFDM systems. Average value estimates of log-likelihood ratio, size of error bursts and probability of error have been compared, but they mostly depend more upon system parameters like number of interferers in CDMA system and subcarrier sizes in OFDM system. HMM performance in the finer resolution, however, could not be established from these data. Comparison of the autocorrelation function and the distribution probability of error-free intervals for the original data and HMM-generated data have been mainly useful to estimate the optimum order of the HMM. The models which have been completely derived for different values of input sample period can be utilized in evaluating of FEC and interleaving schemes for OFDM and CDMA based systems with much less computational burden and simulation time. As compared to realistic physical layer simulation interfaced to the basic OFDM and or CDMA block, the HMM models can substantially reduce computational effort allowing faster, efficient and accurate analysis of higher level protocols and algorithms.

REFERENCES

1. S. M. Z. S. Shah, A. W. Umrani, and A. A. Memon, "Performance Comparison of OFDM, MC-CDMA and OFCDM for 4G Wireless Broadband Access and Beyond", In PIERS Proceedings, pp. 1396 - 1399, Morocco, 2011.

2. H. J. Jeong, and H. Y. Kong, "Performance Analysis of CDMA-OFDM System via Cooperative Communication in Wireless Channel", In 1st International Forum on Strategic Technology, pp. 80-83, Ulsan, Korea, 2006.

3. M. G. Khoshkholgh, K. Navaie, and H. Yanikomeroglu, "Achievable Capacity in Hybrid DS-CDMA/OFDM Spectrum-Sharing", IEEE Transactions on Mobile Computing, vol. 9 (issue 6), pp. 765-777, 2010.

4. T. Balog, F. Balazs, S. Imre and L. Pap, "Analysis of Extended OFDM-CDMA System", In 6th International OFDM-workshop, InOWo 2001, Hamburg, 2001.

5. E. N. Gilbert, "Capacity of a burst-noise channel", Bell System Technical Journal, vol. 39 (issue 9), pp. 1253–1265, 1960.

6. E. O. Elliott, "Estimates of error rates for codes on burst-noise channels", Bell System Technical Journal, vol. 42 (issue 5), pp. 1977–1997, 1963.

7. B. D. Fritchman, "A binary channel characterization using partitioned Markov chains", IEEE Transactions on Information Theory, vol. IT-13 (issue 2), pp. 221–227, 1967.

8. R. H. McCullough, "The binary regenerative channel", Bell System Technical Journal, vol. 47, pp. 1713–1735, 1968.

9. G. Hasslinger, and O. Hohlfeld, "The Gilbert-Elliott Model for Packet Loss in Real Time Services on the Internet", In 14th GI/ITG Conference on Measuring, Modelling and Evaluation of Computer and Communication Systems, pp. 1-15, Dortmund, 2008.

10. G. A. B. Marcondes, and J. M. C. Brito, "Discrete Channel Models Comparison for Transmission over Frequency Nonselective Rayleigh Channels", In 7th International Conference on Mobile and Wireless Communications Networks.

11. L. R. Rabiner, "A Tutorial on Hidden Markov Models and Selected Applications in Speech Recognition", In Proceedings of the IEEE, vol. 77 (issue 2), pp. 257-285, 1989.

12. K. Salamatian, and S. Vaton, "Hidden Markov modeling for network communication channels", In Proceedings of International conference on Measurement and modeling of computer systems, SIGMETRICS 2001, vol. 29 (issue 1), pp. 92-101, Cambridge, Massachusetts, 2001.

13. C. Pimentel, T. H. Falk, and L. Lisboa, "Finite-state Markov modeling of correlated Rician-fading channels", IEEE Transactions on Vehicular Technology, vol. 53 (issue 5), pp. 1491-1501, 2004.

14. P. Sadeghi, Kennedy, R. P. Rapajic, and R. Shams, "Finite-state Markov modeling of fading channels - A survey of principles and applications". IEEE Signal Processing Magazine, vol. 25 (issue 5), pp. 57-80, 2008.

15. Garcia-Armada, B. B. Rodriguez, V. P. G. Jiménez, M. Sánchez-Fernández, "Modelling, Performance Analysis and Design of WPAN

Systems", Wireless Personal Communications, vol. 42 (issue 3), pp. 367-386, 2007.

16. D. Mitra, and R. Ranjan, "Performance Comparison of HMM Discrete Channel Modeling in CDMA Links", International Journal of Computer Applications, vol. 1 (issue 9), pp. 72-77, 2010.
17. Konrad, B. Y. Zhao, A. D. Joseph, and R. Ludwig, "A Markov-Based Channel Model Algorithm for Wireless Networks", Wireless Networks, vol. 9 (issue 3), pp. 189 – 199, 2003.
18. O. S. Salih, C. X. Wang, and D. I. Laurenson, "Three Layered Hidden Markov Models for Binary Digital Wireless Channels", IEEE International Conference on Communications, ICC '09, pp. 1-5, Dresden, 2009.
19. G. Femenias, J. Ramis and L. Carrasco, "Using Two-dimensional Markov models and the effective-capacity approach for cross-layer design in AMC/ARQ-based wireless networks", IEEE Transactions on vehicular technology, vol. 58 (issue 8), pp. 4193-4203, 2009.
20. S. Khudanpur, and P. Narayan, "Order estimation for a special class of hidden Markov sources and binary renewal processes", IEEE Transactions on Information Theory, vol. 48 (issue 6), pp. 1704 – 1713, 2002.
21. A. Akbar, and W. H. Tranter, "Order estimation of binary hidden Markov wireless channel models in Rayleigh fading", In Proceedings of IEEE SoutheastCon 2007, pp. 202-207, Richmond, Virginia, 2007.
22. R. Ranjan, and D. Mitra, "Order estimation of HMM Discrete Channel Model for OFDM systems", In International Conference on Communications, Devices and Intelligent Systems (CODIS'12), pp. 41-44, Kolkata, 2012.
23. Umbert, and P. Diaz, "A radio channel emulator for WCDMA, based on the hidden Markov model (HMM)", In 52nd IEEE-VTS Fall Vehicular Technology Conference, VTC 2000, vol. 5, pp. 2173-2179), Boston, 2000.

24. Ch. V. Verikoukis, "An adaptive hidden Markov model for indoor OFDM based wireless systems", Journal of Telecommunications and Information Technology, vol. 2, pp. 61-65, 2004.

25. R. Kwan, and C. Leung, "An HMM approach to adaptive modulation and coding with multicodes for fading channels", In Canadian Conference on Electrical and Computer Engineering, pp. 147-150, Saskatoon, Canada, 2005.

26. P. Ji, B. Liu, D. Towsley, and J. Kurose, "Modeling Frame-level Errors in GSM Wireless Channels", In IEEE Global Telecommunications Conference, GLOBECOM 02, vol. 3, pp. 2483-2487, Taipei, Taiwan, 2002.

27. W. H. Tranter, K. S. Shanmugan, T. S. Rappaport, and K. L. Kosbar, Principles of Communication Systems Simulation with Wireless Applications. NJ, Prentice Hall, Professional Technical Reference, 2004.

8

A COMPREHENSIVE REVIEW OF SIMULATION TOOLS FOR WIRELESS SENSOR NETWORKS (WSNS)

Anand Nayyar[1], Rajeshwar Singh[2]

[1]Desh Bhagat University, Mandi Gobindgarh, Punjab, India

[2]Doaba Group of Colleges, Nawanshahr, Punjab, India

ABSTRACT

The trend towards adoption of Wireless Sensor Networks is increasing in recent years because of its diverse use in various fields. Wireless Sensor Network is formed via interconnection of large number of sensor nodes. Each and every sensor node deployed in network monitors various parameters like Temperature, Humidity, Ambient Light, Gas etc. and send the data to the master node. Despite of several applications and diverse uses, sensor networks face various shortcomings like energy, localization, security, self-organization, fault tolerance and many more. So, the area of Wireless Sensor Network is under rigorous research and development by various researchers across the globe to develop new algorithms, protocols and techniques to make WSN network more efficient and reliable. Before live implementation, testing of the developed technique requires rigorous testing. But it is not always possible to have live sensor network environment. So, in that case, Simulation is the only way to test the research before moving towards live implementation. Large numbers of simulation tools are available for WSN network till date, out of which some are dedicated towards wireless sensor networks and some for both wireless and wired networks. The main objective behind this research paper is to do a comprehensive review of various simulation tools of Wireless Sensor Networks to enable researcher to select the most competent tool for simulating wireless sensor networks and testing the research proposed. A comprehensive review of 31 WSN simulators is being presented along with their respective features comparison to assist researchers in advanced WSN based research.

KEYWORDS

Wireless Sensor Networks, Simulation Tools, Comparison, Performance Evaluation, Network Simulator

1. INTRODUCTION

In recent years, tons of new and advanced research has been done and is also picking up in the area of Wireless Sensor Networks [1] and this area is also catching the viral attention of researchers from all parts of the world for developing wide applications and making use of WSN networks in varied fields. Wireless Sensor Network comprises of a large number of sensor nodes having sensing and computing capabilities and are deployed in random manner. Each of the sensing devices in WSN network is called MOTE [2].

The following figure shows a view of simple wireless sensor network. Wireless sensor network [4] consists of one or more base stations known as gateways, a number of sensor nodes and end user. The output generated by one node is wirelessly transmitted to the base station for data collection, analysis and logging. Each and every node in the Wireless Sensor Network acts as router for transmitting the information from source node to sink node [122] [123] [125]. The end users are facilitated with the data from the sensor via some website or some application in the console terminal.

Sensor Networks face various kinds of issues which normally don't occur in other sorts of networks [1-4] like power constraints, resource constraints, hardware availability, low communication range, limited processing and storage, cost etc.

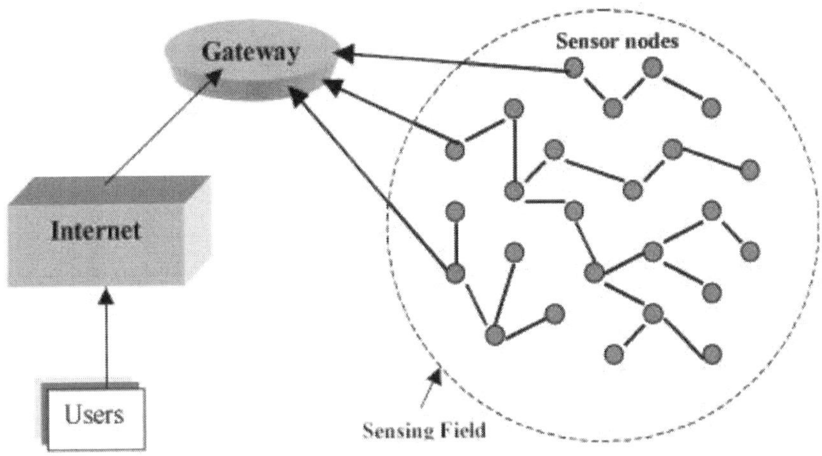

Figure 1. Simple Wireless Sensor Network

As the cost, time and complexity involved in the deployment as well as implementation in such networks is very high, so developers preferably like to get first-hand information on feasibility and reflectivity which is very important for implementation of system before hardware implementation.

So, nowadays various researchers are coming up with new protocols, new techniques and algorithms to overcome the challenges surrounding sensor networks to make it a reliable network and find more possibilities in deploying it in more areas of real world [124]. So, the evaluation and analysis of the techniques being proposed is utmost necessary and desirable. But testing and evaluating the protocols or theories proposed is not practically feasible through real experiments as it would be more complex, time consuming and even costly.

So, to overcome this problem, "SIMULATORS, EMULATORS and TESTBEDS [5] are effective tools to test and analyze the performance of protocols and algorithms proposed. With the help of simulation tools we can easily develop and test a new protocol for sensor networks. There are numerous advantages connected to this approach like easy implementation, low cost, analysis of real time results and even knowing the side effects and other effecting parameters on entire network.

The main objective of this research paper to provide comprehensive review of various Simulators (General Purpose Simulators and WSN Specific Simulators) which are available for researchers to do advanced research on Wireless Sensor Networks.

The Research Paper is organized as: Section II represents Wireless Sensor Network Simulator Architecture, Section III represents Classification criteria for evaluating sensor network simulators, Section IV highlights various Simulators for Wireless Sensor Networks, Section V represents the comparison of various simulators and Section VI presents conclusion and future scope.

2. WIRELESS SENSOR NETWORK (WSN) SIMULATOR ARCHITECTURE

There has been dramatic change in the architecture of WSN simulator from the traditional times to current years. Nowadays, the simulators include new and updated components in terms of energy, power consumption, graphical user interface, and inbuilt protocols cum results interpretation. WSN network is always application specific because the environments in which the sensors are deployed varies from condition to condition. A through test is being required so that to understand the conditions in which the sensor nodes are deployed in order to accurately estimate the node to node interaction [4].

WSN Simulator comprise of following modules: [4] [6]

1) Event

2) Medium

3) Environment

4) Node

5) Transceiver

6) Physical Protocol

7) MAC Protocol

8) Routing Protocol

9) Application Layer

Event: Event lays the primary foundation for functioning of all events and is regarded as abstract class. It comprise of : how event should work, provides procedures to compare events on the basis of their respective start time and whether there is some equality in events and abstract method to fire the event.

Medium: Medium defines the way of communication between the nodes. With medium, the working of nodes is determined. Medium must be informed regarding the presence of nodes, their respective position and other properties like transmitted power or receiver sensitivity. Medium contains the properties of bandwidth and wavelength of medium modelled and a reference to a propagation model at the time of construction. This model will lay the foundation to determine the signal strength transmitted and received by nodes.

Environment: Environment module has similar functionality like medium module. Environment determines the physical properties of model. In terms of WSN network, physical environment include: light, temperature, humidity, sound, optical, coverage area, magnetic field etc.

Node: Node represent the single node in sensor network. As WSN network comprise of various set of nodes which are communicating among themselves via some routing protocol. Node module contains both hardware and software properties of node. Node component includes various properties like: processor, transceiver, sensor, energy source, routing protocol and applications.

Transceiver: This module basically covers the hardware transceiver of sensor node. It covers the transceiver states i.e. Sleep, Standby, Receive and Transmit along with behavior of nodes cum power consumption level of nodes. It generates events for the beginning and ending of every signal it transmits. All these events are exchanged with medium module instance.

Physical Protocol: Being the bottommost layer of network stack, it is mostly implemented in hardware of transceiver. The physical layer provides following services:

- Changing the state of transceiver;

- Carrier Sensing;
- Sending and Receiving Packets;
- Detecting energy on receipt of network packets;
- Channel changing on physical layer which support multiple channel.

MAC Protocol: MAC protocol layer lies between Physical Layer and Routing Layer of network stack and is mostly implemented on software operating in node's processor. In order to provide reliability and efficiency for WSN network the major work is done by MAC layer.

MAC layer is responsible for following services:

- Channel access policies
- Scheduling
- Buffer management
- Error management
- Sending and Receiving of packets between nodes.

In WSN, we need a MAC protocol to consider energy efficiency, reliability, low access delay and high throughput as major priorities to accommodate to sensor's limited resources and to avoid useless power consumption.

Routing Protocol: Routing Protocols comprise of various routing techniques for sending data between sensor nodes and base stations. The Routing Protocols in terms of Wireless Sensor Networks can be classified as follows:

- On the basis of Functioning and Applications: Proactive, Reactive and Hybrid Protocols
- On the basis of Participation Style: Direct Communication, Flat and Clustering Protocols
- On the basis of Network Structure: Hierarchical, Data-Centric and Location Based Protocols.

Application Layer: Application layer comprises of various application-layer protocols which perform various sensor network applications like query dissemination, node localization, time synchronization and network security. This layer is also responsible for sensor nodes traffic management and provides software support for various applications which translate the data in an understandable form or transmit queries to obtain certain information.

3. CLASSIFICATION CRITERIA FOR EVALUATING SENSOR NETWORK SIMULATOR

A. Types of Simulations Surrounding Wireless Sensor Networks

The following are three types of simulation:

- Monte Carlo Simulation
- Trace-Driven Simulation
- Discrete-Event Simulation

Wireless Sensor Network Simulators use Trace-Driven as well as Discrete Event Simulation.

Discrete Event Simulation: Discrete Event Simulation [7] [8] is mostly used in Wireless Sensor Network mainly because of its ease in simulating various jobs running on various sensor nodes. Discrete Event Simulation lists pending events which can be simulated by routines. The global variables lists the simulation time which enables the scheduler for advance time prediction. Simulation includes various routines like input, output, and initial as well as trace. The simulation provides the facility of Dynamic Memory management which facilities the addition and deletion of various entities in simulation along with facility of debugger to check the code side by side without interrupting the normal operation of simulation.

Trace-Driven Simulation: Trace-Driven Simulation [7] mostly plays a crucial role in real time applications. It provides more detailed information which lets the users to have in-depth detail of the simulation model. But side by side trace-driven simulation contains various drawbacks: Sometimes in-depth details increases the complexity of simulation.

B. Evaluation Parameters for Simulators

As lots of simulators are available these days for simulating sensor network. To determine the suitable simulator as per user requirements is not an easy task. So, a criteria [9] [10] [4] to evaluate and compare the simulators is required. The following are some Evaluation Parameters:

1) Type of Simulator: [11] Simulators for Wireless Sensor Networks are classified in following three categories:

   - Generic Simulator: Generic Simulators are especially designed for simulating networking areas. In this simulator, the user creates code in High Level Language which is entirely different as found in real sensor network. Generic level simulator focuses on various aspects of WSN like networking, sensing and data processing. These types of simulators are also used for evaluating new communication protocols but regarded as less reliable due to mismatch in code which is written in simulator as compared to real sensor nodes.

   - Code Level Simulator: Code Level simulators use the same code in simulator as used in real sensor. The code is written and compiled in the machine running simulator which is regarded as faster operation as compared to sensor node. They are OS specific as driver code needs to be replaced when the same code is installed in sensor node. Code Level Simulators are very efficient in locating bugs with regard to timing and hardware architecture.

   - Firmware Level Simulator: Firmware Simulators are based on sensor nodes emulation and software which runs in simulator and is also the same in case of real sensor. Provides more details as compared to generic and code level simulator but side by side, firmware level simulators are also very slow in their operation with regard to code execution.

2) License: Depending on simulator to simulator, some are commercial while some are open source and are available free of cost to download and distribute.

3) Platform: Operating system on which the simulator operates. It can be Windows or Linux while there are some simulators which can operate on both windows and Linux distributions.

4) Usage Popularity: Top Page rank hits in various search engines by searching with "Simulator Name" along with the purpose for which we are looking.

5) WSN Platforms: Various sensors types and platforms which can be simulated in that simulator.

6) Ease of Code: It is also regarded as important criteria for evaluating depending how far the high level language programming is easy to program and how quickly the simulation scenarios can be created with less code and less space. WSN is also bounded to programming languages effectiveness.

Graphical, Debug and Trace Support: (A): Graphical: Whether the simulator has graphical user interface or not; (B) Debugging Aid: How quickly the simulator can detect the error or bug in programming by the user. (C): Trace Support: Representing results in form of graphs. Example: XGraph tool in NS-2 Simulator.

4. SIMULATION TOOLS FOR WIRELESS SENSOR NETWORK

To assist researchers in choosing correct simulator to develop new algorithms, protocols and techniques in Wireless Sensor Network, a detailed explanation of various general purpose and specific purpose simulation tools is presented.

Simulation Tools for Wireless Sensor Network is as follows:

1. NS-2 (Network Simulator-2) [12] [13] [14]

Network Simulator 2 (NS-2) is regarded as discrete event simulation tool and has proved its worth in research of dynamic communication networks. NS-2 was developed in the year 1989 and since its inception various contributions are done that has brought various revolutions in the field of network research. The foundation of this simulator was based on REAL network simulator developed by University of California and Cornell University. Since 1995, DARPA also supported the NS-2 development via VINT (Virtual Inter Network Testbed) and now National Science Foundation (NSF) is also contributing a lot to its development.

NS-2 is totally based on Object Oriented (OO) programming so it is also known as Object Oriented Discrete Event Simulator. It consists of two languages: C++ and Object oriented Tool Command Language (OTcl). C++ is primarily used for implementing various protocols and extending simulation libraries whereas OTcl scripts does the task of configuring simulator, network topology setting, creating network scenarios and displaying simulation results. C++ and OTcl are binded together using TclCL.

NS-2 comprise of 3,00,000 lines of code and is available free of cost and is used globally in academia. It can run on various operating systems like Linux, FreeBSD, MAC OS X, Solaris and even windows via use of third party software called Cygwin. The latest version is 2.36.

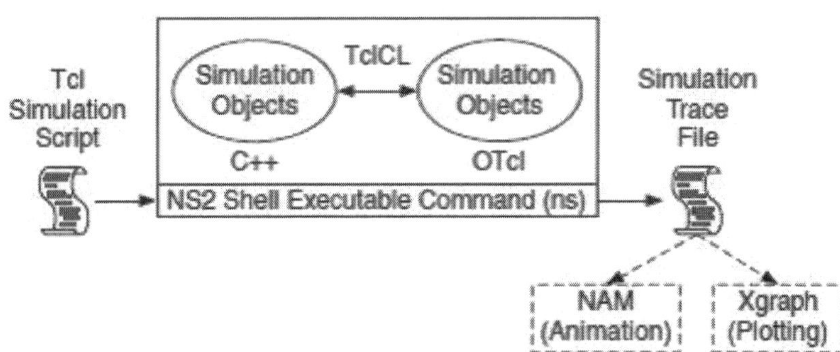

Figure 2. Basic Architecture of NS-2 Simulator [14]

With regard to Wireless Sensor Network, NS-2 provides support for various protocols like 802.11, 802.16, 802.15.4, IR-UWB etc. [9]. But even though of lots of contributions from varied researchers around the world, NS-2 faces serious drawbacks in terms of WSN simulations [5] like: Sensing model doesn't exist. The parameters which are used during simulation of nodes in WSN like energy model, packet formats, and MAC protocols are entirely different as we use in real world sensor network scenario. Another drawback surrounding NS-2 simulation results is that it also lacks application support which is required due to sensor network interaction between application level and protocol level.

NS-2 supports dual output which can be either text-based or graphical based. For graphical based simulation NS-2 has inbuilt tool i.e. NAM (Network Animator) which shows live movement of packets in the nodes, node position and live simulation scenario and also contains XGraphs which shows the graphical analysis of the results drawn at the end of simulation.

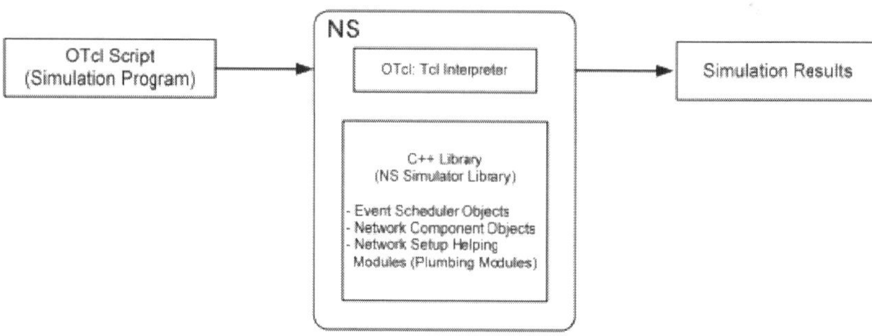

Figure 3. Basic Structure of NS-2 Simulation

2. NS-3 (Network Simulator-3) [15] [42]

Like NS-2, NS-3 is also regarded as discrete-event simulator. The objective behind its development was to enhance research in communication networks. NS-3 is open source simulator and was launched in June 2008. The latest version is 3.21 released in August 2014.

NS-3 is not regarded as extension of NS-2 simulator, NS-3 is a new simulator not supporting any API's belonging to NS-2.

As programs written in NS-2 are coded in OTcl and results can be visualized using NAM and XGraph but pure C++ code is not possible in NS-2. But in NS-3, all the programs are written in pure C++ with optional python bindings. There is no Graphical Tool which is available in NS-3, but still graphical results can be interpreted using NetAnim open source software.

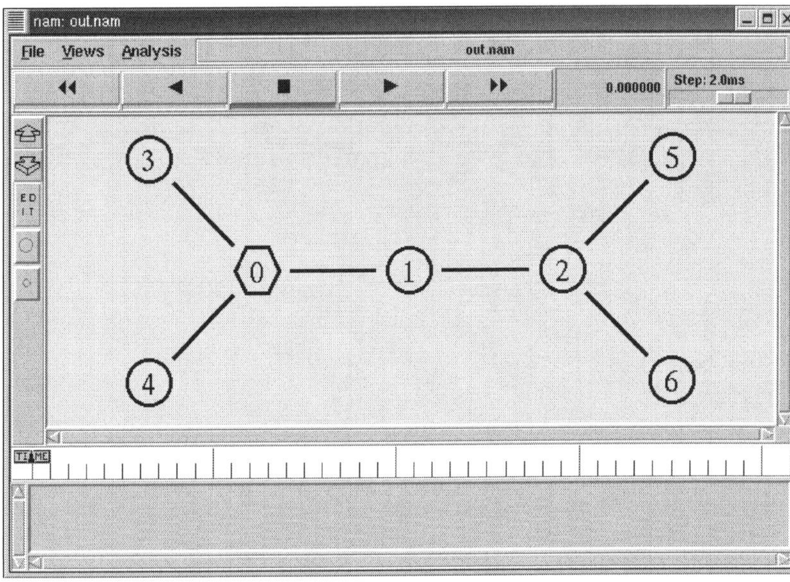

Figure 4. NAM (Network Animator in NS-2)

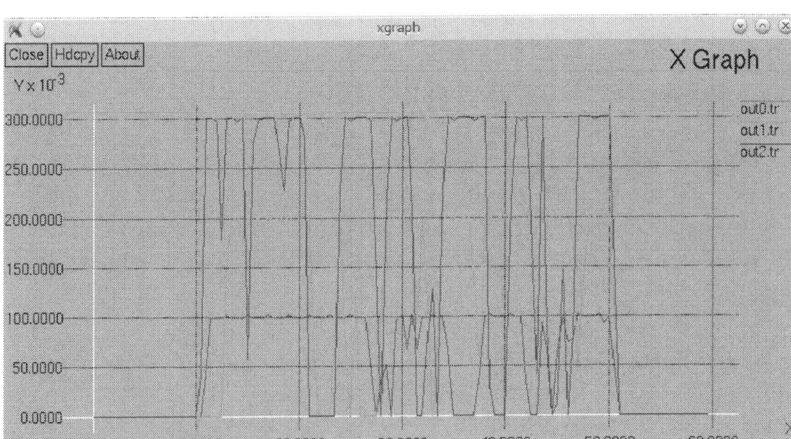

Figure 5. XGraph Tool in NS-2

For wired topology, NS-3 provides device model of a simple network of Ethernet which uses CSMA/CD as its protocol scheme with exponentially increasing back-off to contend for the shared transmission medium. With regard to Wireless Sensor Networks, various modules like 802.15.4, 6LoWPAN and RPL are already integrated in NS-3.

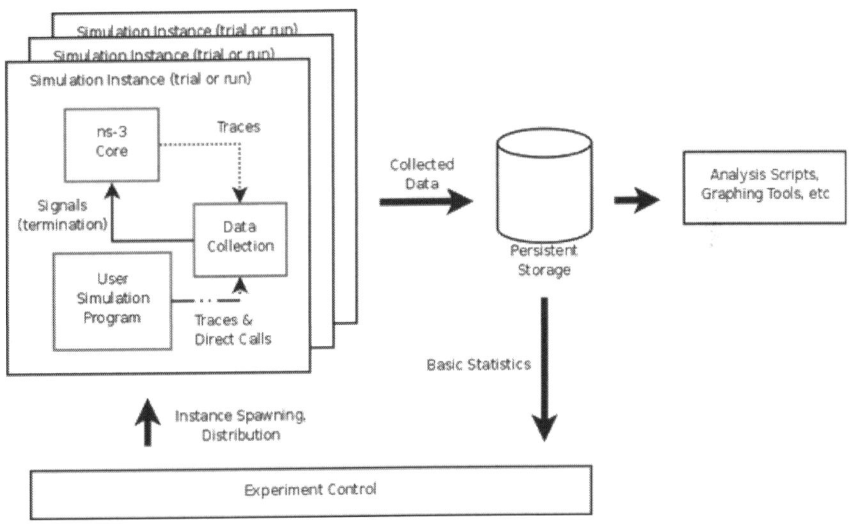

Figure 6. Architecture of NS-3 Simulator

NS-3 can also work in various operating system platforms like Linux and Windows via Cygwin.

3. OMNeT++ [16] [9] [5] [4] [17] [20]

OMNet++ [16] is another powerful object oriented discrete event network simulator for wireless sensor networks. It is regarded as extensible, modular and component based C++ simulation library framework for building wireless network simulations. Basically OMNeT++ is not a simulator but it provides frameworks and tools for writing simulation scenarios. It is compatible to run on different operating systems like Windows, Linux and MAC OS X. The current version of OMNeT++ is 4.6 which was released in July 2014 with primarily focused on improving the usage of Graphical Runtime Environment (TKenv). TKenv GUI is re-engineered for single-window mode to improve the user experience and usage of simulator. The primary role of TKenv is to show the animation of network model, movement of nodes, displaying results, visualization changes etc. OMNeT++ is free to download for education and research purposes. Company has also got licensed version called "OMNEST" and license is available via Simulcraft Inc.

The basic elements which make up OMNeT++ are called modules. Modules are of three types: (A): Simple Module: It is written in C++; (B): Compound Module: The module which links other modules via connections; (C): Network Module: Top level compound module is called network module.

OMNeT++ includes Eclipse based IDE environment which enables C++ programming and debugging of modules as well as graphical and textual editing of NED Files. NED (Network Description) lets the user declare simple modules and connect and assemble them into compound modules.

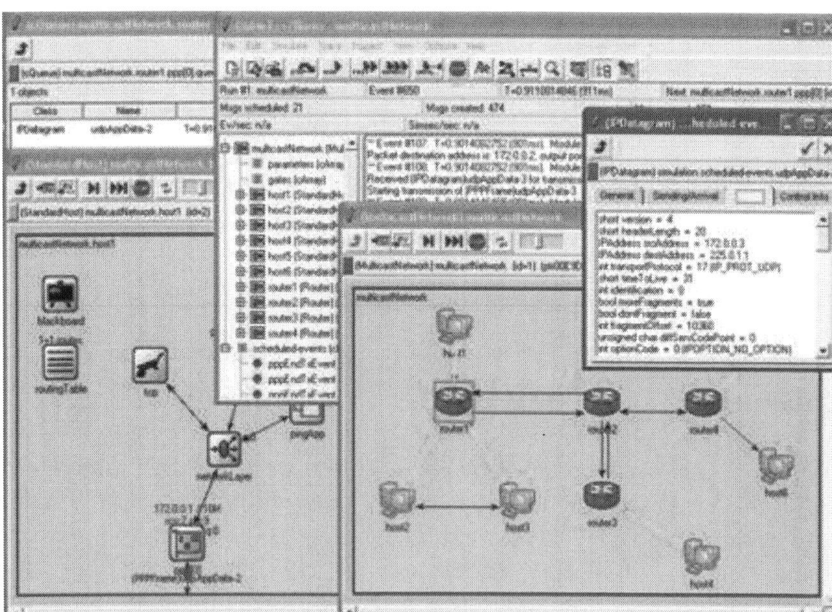

Figure 7. OMNeT++ Simulator IDE Interface

The following is the basic NED code which is being written to make the topology in OMNeT++:

```
//
// A network
//
network Network
{
    submodules:
        node1: Node;
        node2: Node;
        node3: Node;
        ...
    connections:
        node1.port++ <--> {datarate=100Mbps;} <-->
node2.port++;
        node2.port++ <--> {datarate=100Mbps;} <-->
node4.port++;
        node4.port++ <--> {datarate=100Mbps;} <-->
node6.port++;
        ...
}
```

NED Code for Topology Description

The Simulator which is built primarily for Wireless Sensor Network in OMNeT++ is CASTALIA.

CASTALIA [18] [19]

Castalia, a popular Wireless Sensor Network, Body Area Networks (BAN) simulator which is made especially for low power embedded devices. The simulator is primarily being used by researchers to simulate algorithms and protocols in real time wireless channels and radio models and to see the real time node results and study interpretations.

Castalia is not regarded as sensor specific platform rather it was developed to provide real time results of the algorithm being developed by the researcher on any specific sensor platform. The latest version of Castalia is 3.2 which was released in March 2011.

Castalia provides the following:

- Channel Model to study mobility of nodes, study the path between the nodes and any interference level with regard to signal strength.
- Radio Model to interpret SINR, packet size, module type and carrier sensing.
- MAC and Routing Protocols study on nodes.
- Monitoring CPU power consumption.

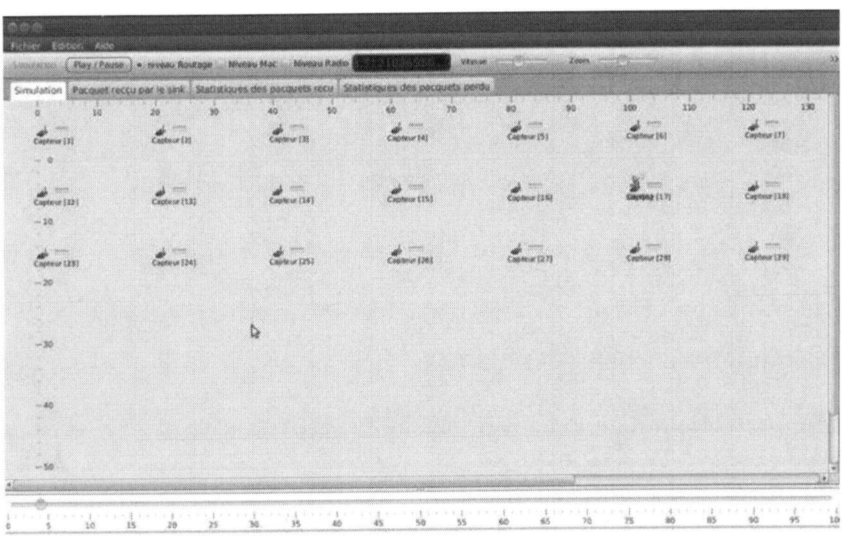

Figure 8. Castalia GUI for WSN Simulation

4. J-Sim [21] [22] [23] [24] [25] [5] [17]

J-Sim (JavaSim) Simulator is regarded as general purpose simulator and is based on component based software architecture, Autonomous component architecture (ACA). J-Sim has been developed in Java and is tightly coupled with ACA making J-Sim a platform neutral, extensible and reusable environment. J-Sim has a Script Interface which facilities execution of various script based languages like Perl, TCL or Python. The latest version of J-Sim is 1.3 + patch 4 which facilitates J-Sim with a

Java implementation of TCL interpreter with TCL/Java extension known as Jacl.

J-Sim has a sharp edge over NS-2 simulator as unlike in NS-2, classes/methods/fields in Java need not be explicitly exported in order to be accessed in TCL environment; rather all the public classes/methods/fields in Java can be accessed in TCL environment.

J-Sim consists of the following packages:

- Base Package
- NET Package
- INET Package
- Diffserv Package
- Intserv Package
- RUV System
- Common Class Library
- J-Sim Total
- Third Party which consists of: Jacl; Jython; PtPlot

The following figure shows the GUI of J-Sim Simulator:

J-Sim provides a framework for Wireless Sensor Network Simulation which has been built upon ACA and INET and Wireless Protocol stack and has following components:

a) Target, Sensor and Sink Nodes.

b) Sensor Channels and Wireless Communication channels

c) Physical media such as Seismic channels, mobility models and power models.

Considering simulation of Wireless Sensor Networks, J-Sim has used to implement three protocols of WSNs: Localization, Geographic Routing and Directed Diffusion [24]. As compared to NS-2, J-Sim simulator has various advantages in terms of execution time, memory allocation, and scalability. Moreover it provides good platform to carry out large-scale WSN simulations with more than 1000 nodes which is somewhat not possible in NS-2. J-Sim provides framework for several classes for simulating sensor network in terms of coverage, time indexing and power management.

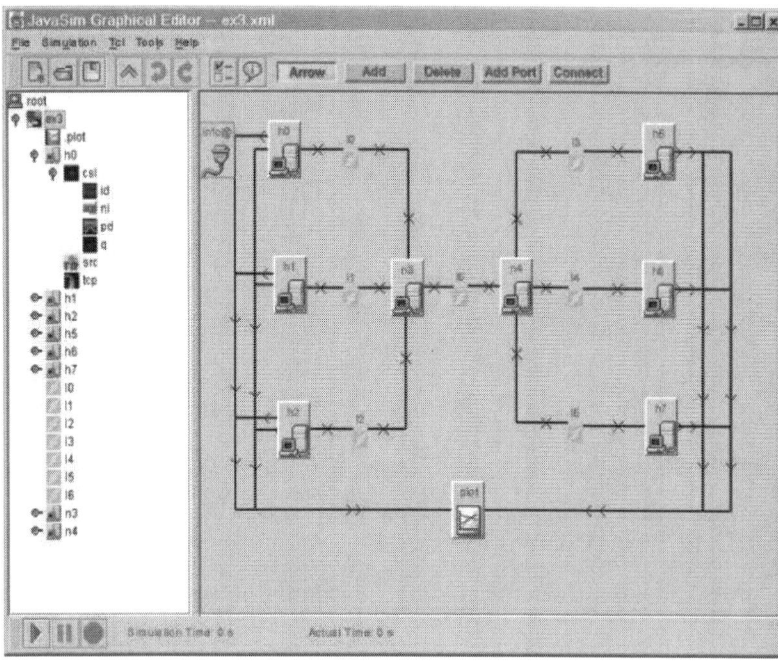

Figure 9. J-Sim GUI

5. Mannasim [26] [27]

Mannasim Simulator is basically a module for WSN simulation based on Network Simulator 2 (NS-2).

Mannasim, a wireless sensor network simulator is comprised of two components:

a) Mannasim Framework

b) Script Generator Tool

Mannasim Framework: Mannasim framework is basically an extension of NS-2 simulator and provides new modules for design, development and analysis of various Sensor network applications.

Script Generator Tool (SGT): Mannasim Script Generator Tool provides easy to use front end for creation of TCL simulation scripts. It is part of Mannasim framework and is fully written in Java Language making it platform independent.

The following figure highlights SGT tool of Mannasim:

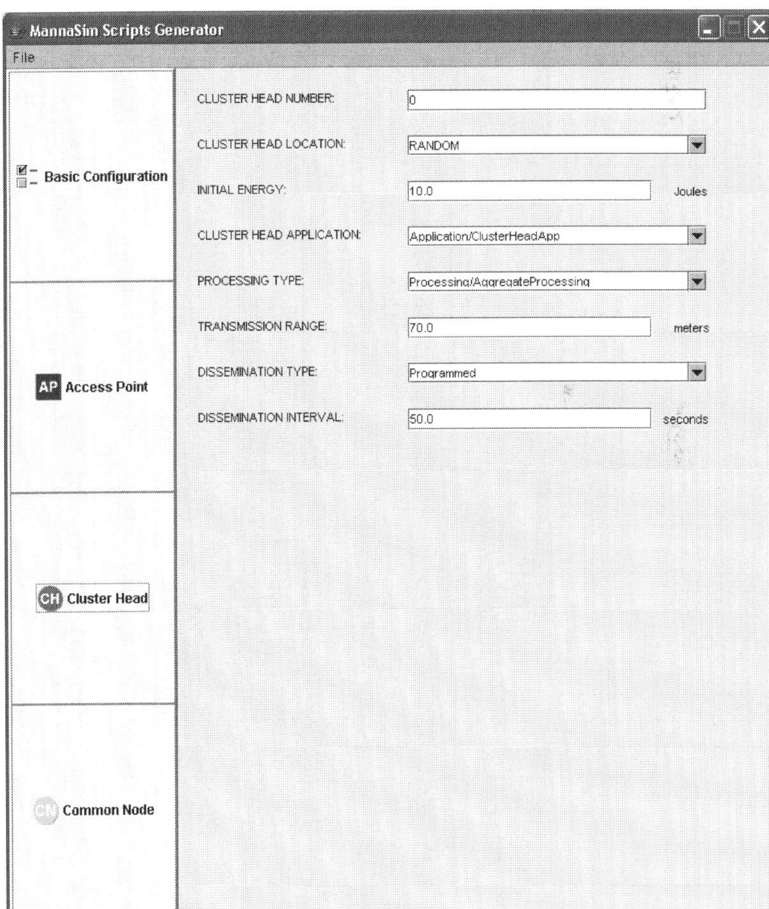

Figure 10. Mannasim Script Generator Tool

In case of Wireless Sensor Network scenario, Mannasim provides a detailed simulation framework for accurate modeling of sensor nodes and comprehensive testbed for simulating various algorithms and protocols of Wireless Sensor Networks.

It is regarded as highly signified tool for WSN to simulate the nodes and analyze the performance of nodes on various algorithms. Various WSN routing protocols like AODV, DD, DSDV, DSR, LEACH, and TORA can be simulated.

Mannasim can be installed along with ns-2.29 patch and is open source.

6. SensorSim [5] [28] [29] [30]

SensorSim, an another framework for simulating wireless sensor network is based on NS-2 simulator with various add on features like: (a) Sensing channels and Sensor Models (b) Battery Models (c) Lightweight protocol stacks for wireless microsensors (d) Scenario Generation (e) Hybrid Simulation.

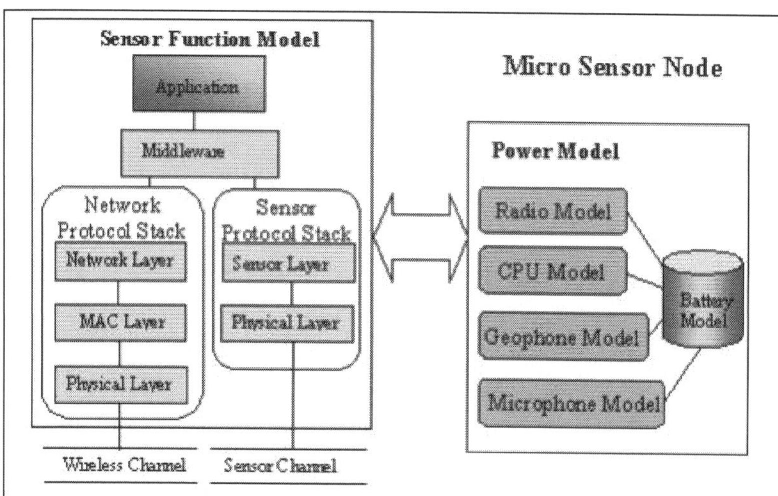

Figure 11. MicroSensor Node Model

The basic objective behind the development of SensorSim is to provide deeper understanding of sensor networks and to provide a deep framework for development of new protocols cum applications and performance evaluation techniques for sensor networks. SensorSim has inherited various features from other event driven simulators having hybrid simulation mode which facilitates new features like interaction of real and simulated nodes, new communication protocols and real time user interaction with GUI interface.

Apart from various advantages and new features for simulating WSN network, SensorSim has scalability problem. And due to inability of the developers to provide support for public release, SensorSim today is not currently available to the public.

7. NRL Sensorsim [13] [31] [32] [27] [33]

NRL's Sensorsim is basically an extension to NS-2 simulator for simulating Wireless Sensor Network and was developed by The Naval Research Laboratory group. NRL Sensorsim supports simulation and detection of Carbon Monoxide concentration, Seismic activity and audible sound. NRL Sensorsim also has the feature to support energy consumption model in sensor network. NRL Sensorsim is now no longer in development and no support is being provided to public.

The following diagram shows the GUI of NRL Sensorsim:

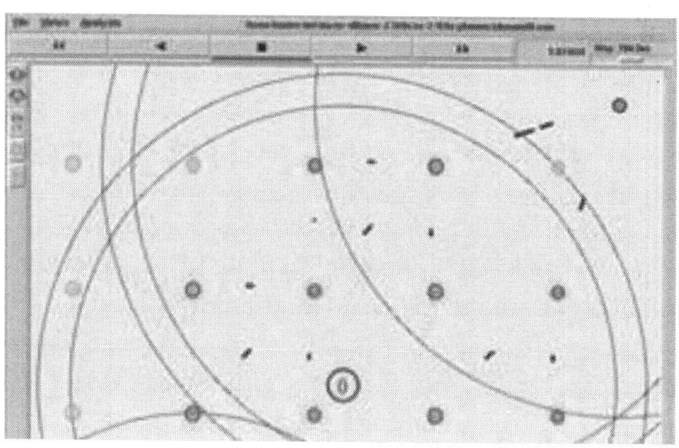

Figure 12. NRL Sensorsim GUI

8. NCTUns 6.0 [6] [34] [35] [36] [37] [38]

NCTUns- (National Chiao Tung University Network Simulator) - An Open Source Discrete event network simulator cum emulator software runs on Linux and provides easy to use integrated GUI environment for network planning and research.

The main components of NCTUns Simulator are: GUI, job dispatcher, simulation engine, patched Linux kernel, applications and daemons.

- GUI provides user interface to create and control the simulation in an interactive manner.
- Job dispatcher monitors the status of various machines which are participating by receiving status reports sent from coordinator programs running on them.
- Coordinator Program monitors the machine status that whether the machine is busy or idle and in turn send the report to Job dispatcher.
- The following figure shows the architecture of complete NCTUns Simulator:

The latest version release of NCTUns is 6.0 which was released in May 2010.

The following are the features of NCTUns Simulator:

a) Uses real world Linux TCP/IP protocol stack to generate high end simulation results.

b) Can use any existing or in-development UNIX program as Traffic Generator program and needs no modification.

c) Can use real world UNIX network configuration and monitoring tools.

d) Simulation environment is almost the same as Real world IP networks.

e) Facilitates simulation of various networks like WSN, MANETS, Optical networks, GPRS, DVB/RCS Satellite network, WiMAX network, IEEE 802.11p/1609 VANETS.

f) Can simulate almost every network device as per IEEE standards may be Wired or Wireless device.

g) Has ability to simulate large quantity network protocols like CSMA/CS, HTTP, FTP, RIP, OSPF, 802.11 etc. to name a few.

h) High simulation speed

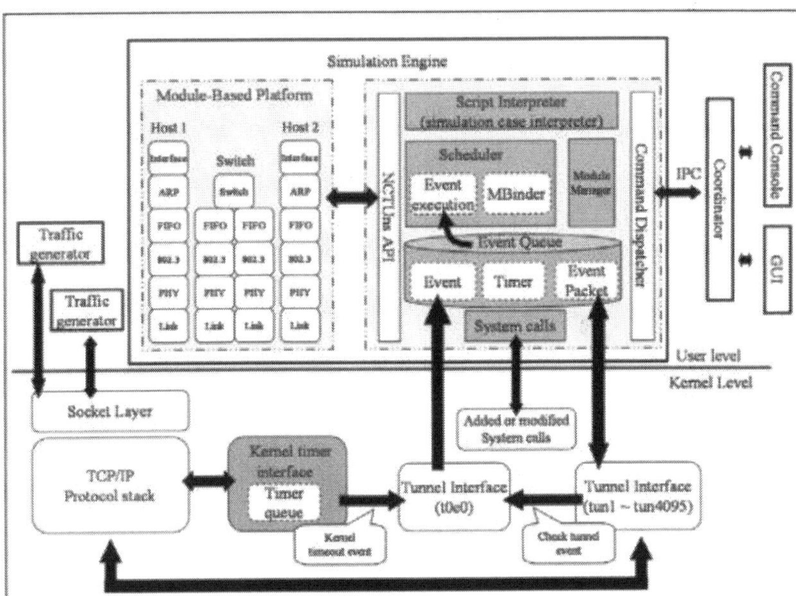

Figure 13. NCTUns Simulator Architecture

The following figure shows the GUI interface of NCTUns Simulator.

With regard to WSN Simulation, NCTUns Simulator efficiently simulates various routing protocols like AODV, DSR, and DSDV.

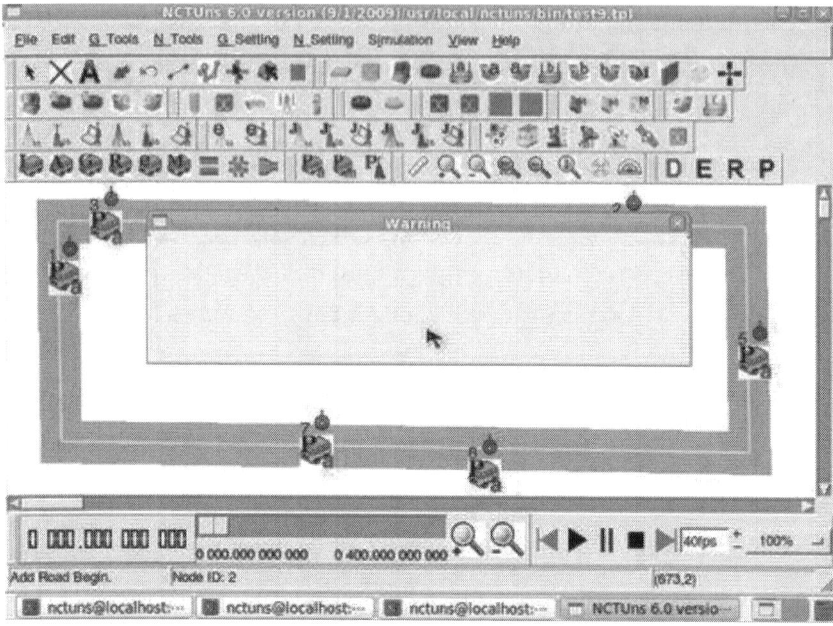

Figure 14. GUI interface of NCTUns Simulator

9. SSFNet [6] [39] [40] [41] [42]

SSFNet-Scalable Simulation Framework Network Models is an open source simulation software for simulating networks especially Wireless Sensor Networks. It is basically designed for simulating various network scenarios like network topology, protocols and traffic and also enables simulation of Wide Area Network like Internet.

SSFNet is a complete Java based implementation of SSF engine along with various highly scalable Internet protocols like IP, TCP, UDP, BGP, OSPF etc. and large network elements like Routers, Switches, Links, LAN's etc.

SSFNet models are regarded as "Self-Configuring" which means each SSFNet class instance can configure itself separately by querying the parameters in the database which may be local or over the web.

SSFNet models provide components for simulating networks at IP level and above and include models for hosts, router, links and a framework for modeling protocols.

The following figure shows the Simulation Layers of SSFNet Simulator:

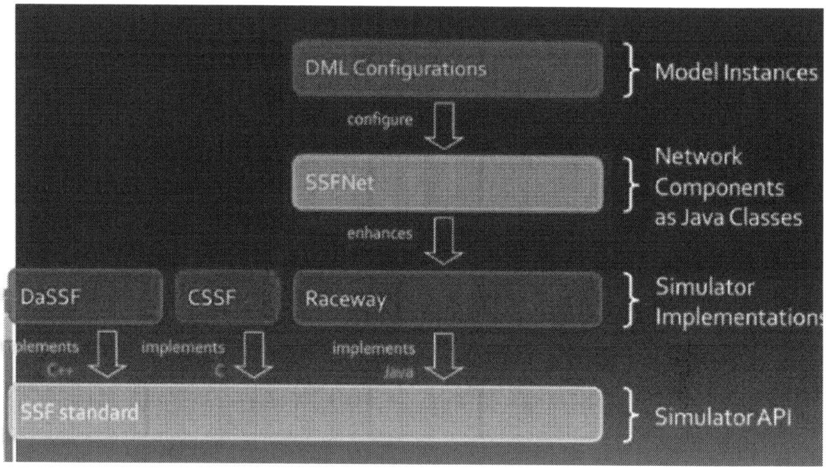

Figure 15. Simulation Layers of SSFNet

SSFNet is basically divided into the following four packages:

a) SSF.OS- It contains the framework for protocols models that are related to operating system.

b) SSF.Net- It includes the hardware for modelling network objects such as hosts, routers, links etc.

c) SSF.Util.Random- It contains additional classes for generation random stream.

d) SSF.Util. Streams- It contains the support for efficient multiple-point network, monitor the infrastructure for collecting stream data.

The current release of SSFNet simulator is 2.0 was done on January 15, 2004.

10. GloMoSim [42] [43] [44] [17] [4] [45] [5]

GloMoSim (Global Mobile Information System Simulator) is discrete event scalable simulation software that simulates wireless and wired network systems and is built using PARSEC simulation environment. GloMoSim as well as PARSEC were developed by Parallel Computing Lab at UCLA.

A layered approach is being used by GloMoSim to carry out simulation between different layers and a Standard API is being used so that new models and layers can be exchanged and integration can be facilitated.

Two different node mobility models: Random Waypoint Model and Random Drunken Model is being supported by GloMoSim. When the nodes in the network move according to a model, it is known as Random Waypoint model.

When a node in the simulation environment chooses random destination and moves to the location as per the speed specified in configuration file it is known as Random Drunken Model.

The following Table 1 [44] shows GloMoSim models at major layers of OSI model:

Table 1. Models in GloMoSim Simulator

Layer	Model
Physical Layer	Free Space, Rayleigh, Ricean, SIRCIM
Data Link Layer	CSMA, MACA, MACAW, FAMA, 802.11
Network Layer	OSPF, DSR, WRP
Transport	TCP, UDP
Application	Telnet, FTP

GloMoSim was non-commercial simulator and the current version which is available these days is version 2.03. But now a Commercial version of GloMoSim has been evolved with name "*QualNet*".

With regard to Wireless Sensor Networks, basic routing protocols like AODV, DSR is available in it. But the current version of GloMoSim doesn't offer any sensor network features. But GloMoSim provides good platform for simulating Mobile IP Networks.

The following figure shows the GUI interface of GloMoSim simulator:

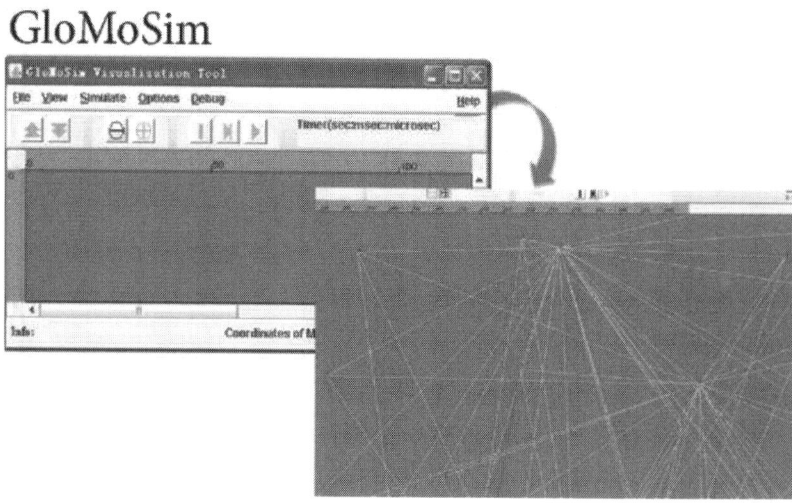

Figure 16. GUI Interface of GloMoSim Simulator

The following figure highlights the architecture of GloMoSim Simulator:

PARSEC (PARallel Simulation Environment for Complex Systems) [46] [44].

PARSEC (PARallel Simulation Environment for Complex Systems) was designed by Parallel Computing Laboratory, UCLA. It is basically a C-Based simulation language for sequential and parallel execution of discrete-event simulation models.

PARSEC supports both Windows and UNIX platforms. PARSEC is based on Maisie Simulation language with following major modifications:

- Simple Syntax.
- Usage of modified language for posting code from simulation model to operational software.

- Robust and Efficient runtime kernel.
- Integration of new protocols for parallel performance prediction.
- The following figure gives a visual representation of PARSEC Environment:

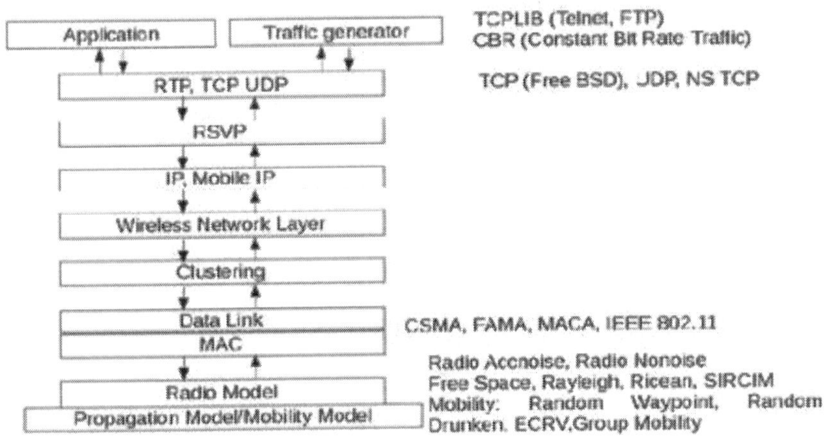

Figure 17. GloMoSim Architecture

11. QualNet 7.0 and EXata 5 [42] [17] [48]

QualNet-(Quality Networking) a commercial network simulation software from Scalable Technologies is basically derivate product from GloMoSim Simulator. QualNet enables high fidelity "Virtual" model of network which consists of routers, switches, servers, access points, radios, antennas, computers and any other equipment's along with various protocols to enable packet movements across the network scenario.

QualNet 7 and EXata network simulator is supported across various platforms like Windows 7/8/8.1, Centos 5.9, RHEL 6/7 and Ubuntu 12.04 version onwards and was released in October 2013.

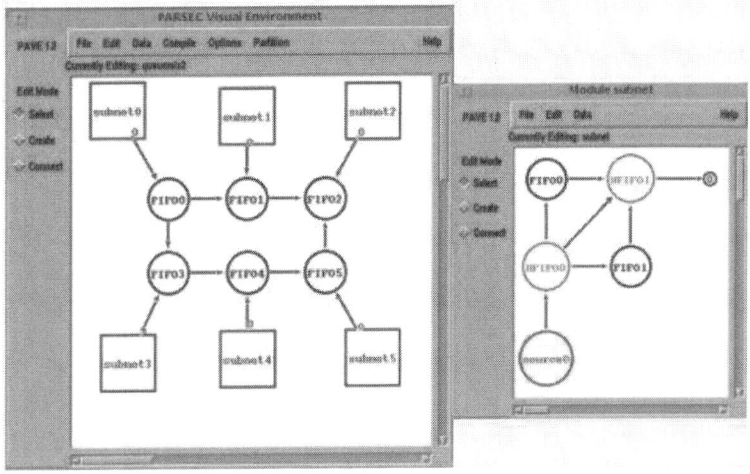

Figure 18. PARSEC Visual Environment GUI

The following figure shows the GUI interface of QualNet 7 Simulator:

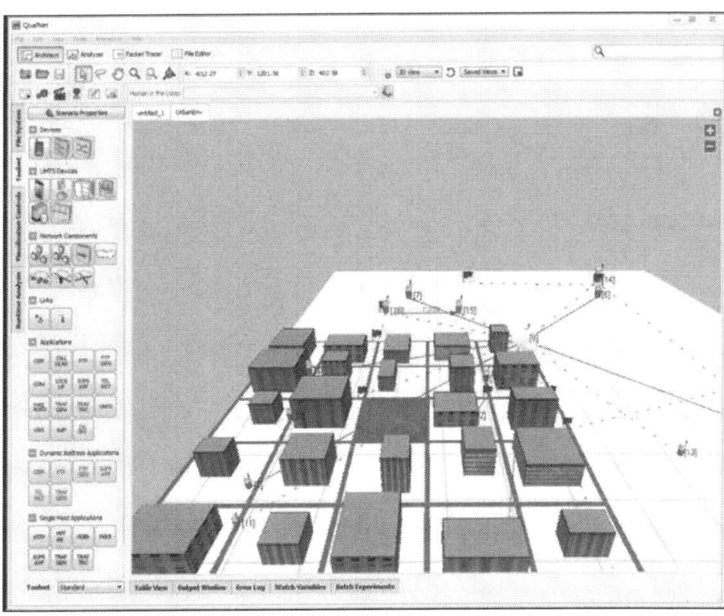

Figure 19. GUI Interface of Qualnet 7 Simulator from Scalable-Technologies

The following Table 2 highlights the components of QualNet Simulator:

Table 2. Components of QualNet Simulator

Component	Description
1. Architect	A GUI Design and Visualization tool to setup network scenarios, connections, users and even nodes. Architect tool enables in-depth visualization and network topology analysis. It also enables showing of real-time statistics via graphs depending on the simulation running.
2. Analyzer	A Graphical Statistical tool used for displaying various parameters of simulation undergoing on. All the statistics can also be exported to MS-Excel via CSV format.
3. Packet Tracer	GUI Tool for showing visual representation of packet trace files which are generated during network simulation. Basically text files in XML form and shows the details of packet movements across various layers of OSI model.
4. File Editor	Tool basically used for text editing.
5. Command Line Interface	Enables CUI interface to the simulator.

The following figure shows QualNet Layer Model.

EXata: A simulation platform which contains all functionality of QualNet with a real-time network emulation interface which enables the users to integrate hardware and run applications on the simulation environment. EXata enables live simulation of cyber vulnerabilities and threats so that research can be undertaken by users to prevent network security against outside attacks.

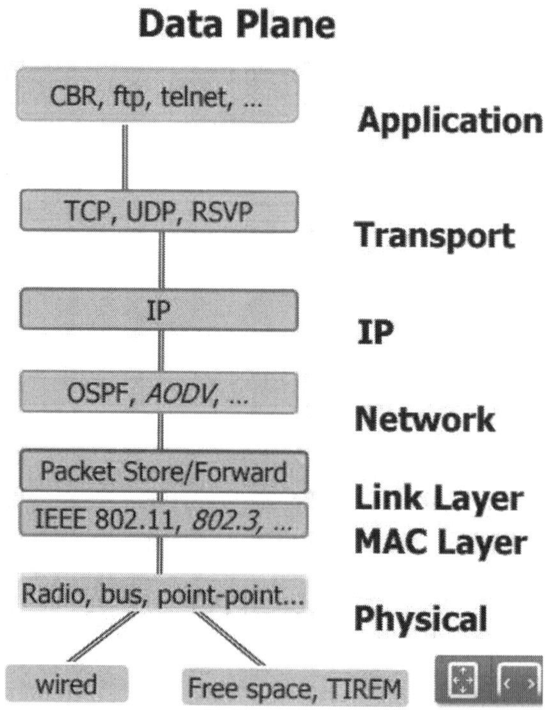

Figure 20. QualNet Layer Model

The following figure demonstrates the GUI interface of EXata Simulator.

The following are the features which are included in QualNet 7 and EXata 5 making them as the most advanced simulator platform these days:

- IPv6 Emulation Support
- IGMPv3 Support
- DHCP Model
- DNS Model
- FQDN (Fully Qualified Domain Name) in Applications
- IPv6 Auto configuration Model

- Channel Names
- Federate Extractor
- HLA 1516
- RPR FOM 2.0017

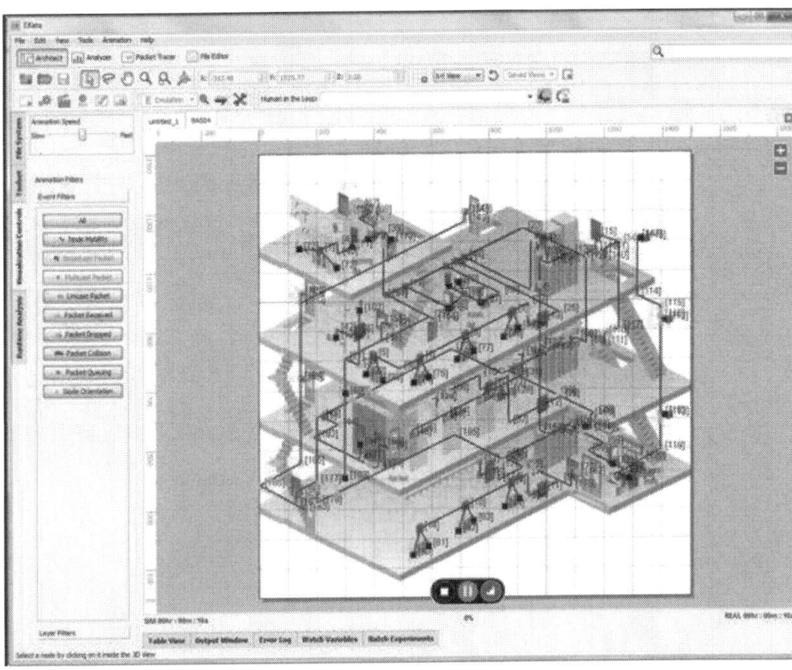

Figure 21. GUI Interface of EXata Simulator

12. sQualNet Simulator [49-52]

sQualNet-Scalable Quality Networking a scalable network simulator based on QualNet. sQualnet has wide suite of sensor network specific models i.e. Sensing and Radio Channels, Sensor Protocols (MAC, Routing), battery and power consumption models along with support for multi-tiered sensor network evaluation. sQualnet has an added edge over various other simulators like NS-2, OPNET, QualNet and other sensor specific simulators is that it has ability for real-code simulation.

It also has advantages over emulators like TOSSIM, ATEMU in terms of hybrid simulation which provides assistance in large scale system development and evaluation at low cost.

It is also regarded as kernel efficient protocol which allows experiments to run at higher scalability in both contexts i.e. Spatial and Temporal.

The latest version of sQualnet available currently is 1.0 Beta. It was a project funded by National Science Foundation (NSF) and Defense Advanced Research Projects Agency (DARPA) with joint collaboration with Network and Embedded Systems Laboratory (NESL) and Parallel Computing Laboratory at University of California, Los Angeles (UCLA).

The following figure shows the Capabilities of sQualnet Simulator:

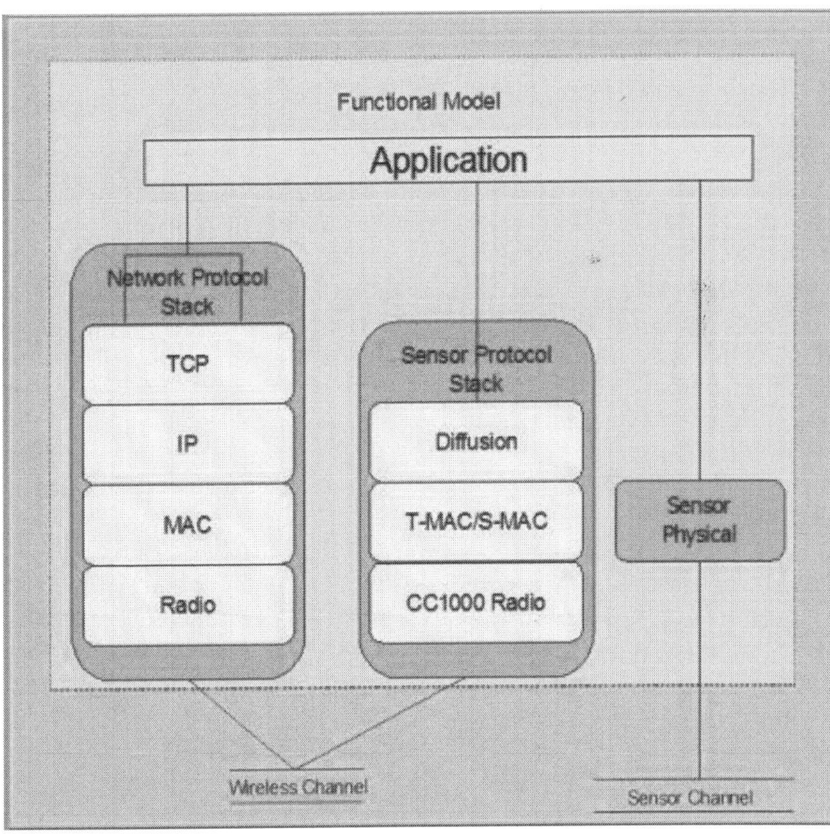

Figure 22. sQualnet Simulator Capabilities

13. OPNET Modeler Suite [53-60] [42] [17]

OPNET (Optimized Network Engineering Tools) was created by OPNET technologies in 1986. The Latest version of OPNET Modeler Suite is 18.0.3 which was recently released in April 2015. Now, OPNET is part of Riverbed Application and Network Performance Management Solutions.

OPNET tool is a powerful collection of tool sets to create and test large network environments because of its coverage to following components of communication networks:

a) Application Performance Management

b) Planning

c) Engineering

d) Operations

e) Research and Development

OPNET provides an interactive GUI interface to build network scenarios and enables its testing. Basically users can make use of drag and drop actions to create high end network environments of any sort in matter of seconds. OPNET Modeler and IT Guru are basically the packages used for network simulation.

OPNET Modeler is basically a commercial version available but IT Guru Academic Edition is basically a free software license given by OPNET as part of University Program throughout the world.

The following figure highlights the GUI interface of OPNET Modeler 18.0.3.

OPNET is a discrete event simulator and is available both in 32-bit as well as 64-bit for Windows and Linux Operating System. OPNET Riverbed modeler makes use of Object Oriented technology to create mapping from the graphical design to the implementation of the real system.

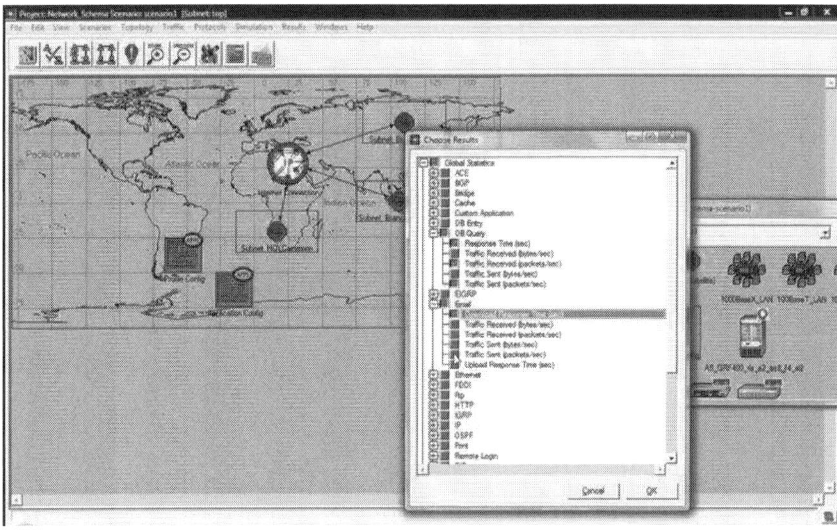

Figure 23. OPNET Modeler Interface

OPNET models the network in a hierarchical approach which somewhat matches to hierarchical structure of Internet: Networks, Satellites and Nodes. OPNET basically supports three types of Links: Point-to-Point, Bus and Wireless. Wireless Link is basically used in Wireless, Mobile or Satellite Network. Although, OPNET has various libraries for Wireless and Wired networks, it also supports Wireless Sensor Networks. OPNET basically comprise of three main tools: Net Doctor, ACE and MVI which enables the analysis of networks and proper future implementations.

OPNET 18.0.3 Modeler provides various editors to facilitate easy network simulation.

The following Table 3 shows the list as well as description of various editors in OPNET.

14. SENSE [5] [24] [8] [27] [17] [61] [62] [63] [71]

SENSE: Sensor Network Simulator and Emulator a powerful sensor network simulator developed in 2004. The latest version available for SENSE is 3.0.3.

Table 3. List of Editors in OPNET Simulator

Editor Name	Description
Project Editor	Used for main staging for network simulation. Basically network model is built here using models from library, statistics choosing about network, simulation running and results viewing. We can create and process nodes, build packet formats, create filters and parameters in Project Editor.
Node Editor	Basically helps the user in defining the behavior of each network object. Network Object comprise of multiple modules which defines the behavior of node. Modules comprise of data creation, data storage etc.
Process Model Editor	It helps to create process models which are represented via Finite State Machines (FSMs). In this operations are being performed via C/C++ code.
Link Model Editor	It creates new types of Link objects which can have different attributes and representations.
Path Editor	It is used to create new path objects which define a traffic route. Logical Connections like MPLS, ATM or Frame relay are being used as path for traffic routing.
Demand Editor	It is used to define demand models and it determines attributes interfaces, presentation and behavior.
Probe Editor	It generates statistics of the simulation.
Simulation Sequence Editor	It helps to specify additional simulation constraints and contains a set of various attributes which control runtime simulation.
Antenna Pattern Editor	It controls the direction-dependent gain properties of antenna. This editor is only used in Wireless Simulations.
Filter Editor	It creates additional filters to build new filter models.
ICI (Interface Control Information) Editor	It defines the internal structure and used for interrupt-based inter-process communication.
Modulation Curve Editor	It plots BER and SNR.
Packet Format Editor	It defines internal structure of packet as set of fields.
PDF (Probability Density Function) Editor	It defines overall results and any sort of transmission errors.
Project Editor Window	Basically a GUI window for defining simulation and user can drag and drop components of various network devices on this window.

SENSE is a discrete event simulator written in C++ and its working is similar to J-Sim and runs on the top of COST, written in CompC++ (A component based extension to C++). The design of COST was basically influenced by mutual understanding of Component-based software architecture and component-based simulation.

With regard to Wireless Sensor Networks, SENSE has battery models, MAC Layer protocols like NullMAC, IEEE 802.11 with DCF, application, network and physical layer models. SENSE also includes a series of SSR Protocols like: Self-Selective Routing (SSR), Self-Healing Routing (SHR) and Self-Selective Reliable Path (SRP).

For results visualization and simulation animation, SENSE has iNSpect tool. The latest version of iNSpect tool is: 3.5.

The most important critical factors of SENSE are:

a) Extensibility: Network simulation architecture is based on component based simulation. Component-Port model removes interdependency in simulation models as in OOPS architecture and then a simulation component classification is being used to solve the problem of managing simulation time.

b) Reusability: Removal of interdependency between models lays foundation of reusability. In SENSE simulator, there is another level of reusability which was made possible by extensive use of C++ template: a component is usually declared as a template class so that it can handle different type of data.

c) Scalability: Like SSFNet and GloMoSim, SENSE also includes the support for parallelization.

The following Table 4 shows the list of various components and simulation engines available right now in SENSE Simulator.

The following figure shows the GUI interface of G-Sense Visualization Tool for Simulating AODV Protocol.

SENSE, is still in development phase and developers are releasing new patches for its stability but still lots of new models and templates are to be developed with regard to Wireless Sensor Networks.

Table 4. Components of SENSE Simulator

Component Name	Short Description/Ingredients
G-Sense [64] [65]	Guide User Interface for SENSE simulator especially for WSN simulation and includes various simulation features like parameter input, stop time, Number of nodes, Terrain Size, Number of Source Nodes, Packet Size and Interval.
Battery Model	Linear Battery; Discharge Rate Dependent
Application Layer	Random Neighbor, Constant Bit Rate
Network Layer	Simple Flooding, AODV, DSR, SSR, SHR
MAC Layer	NullMAC, IEEE 802.11 with DCF
Physical Layer	Duplex Transceiver
Wireless Channel	Free Space, Adjacency Matrix
Simulation Engine	CostSimEng (Sequential)

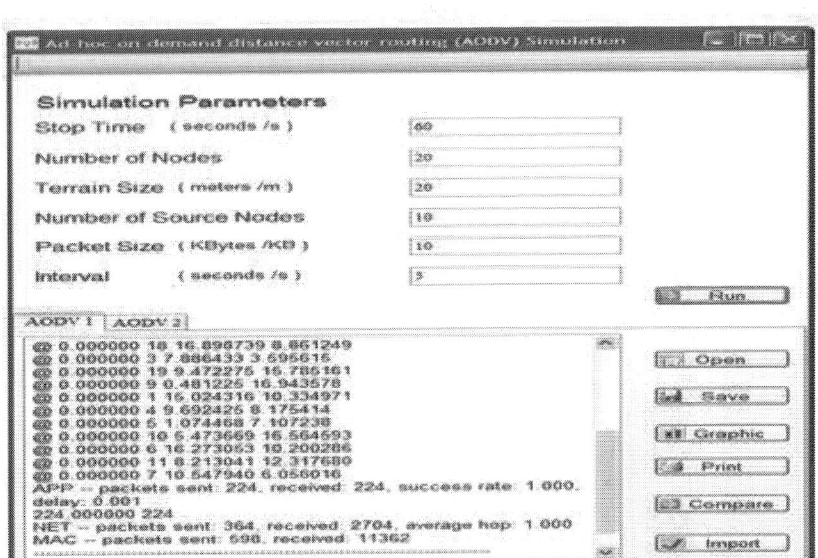

Figure 24. AODV Simulation in G-Sense Simulator

15. DRMSim [42] [66] [67] [68] [69] [70] [71]

DRMSim-Dynamic Routing Model Simulator- A Discrete event simulation software, a joint research project venture by Alcatel-Lucent Bell, Universitée de Bordeaux (LaBri) and INRIA at Sophia Antipolis (Mascotte project) and is supported by Euler fire project whose main is to research and develop a new routing scheme suitable for future Internet.

As basic routing protocol i.e. BGP (Border Gateway Protocol) which drives today's Internet shows various limitations in terms of routing table entries which are stored locally, updated in timely fashion and dynamic exchange. It is right now somewhat practically impossible to deploy new-designed protocols on Internet at a very high-scale so before deployment, a valid simulation is prime requirement. So, the idea behind the development of DRMSim was to simulate routing models on large-scale networks. DRMSim enables simulation of large-scale network with minimum of 10000 nodes which in turn requires careful analysis of data structures that will be used on the network model as well on the granularity and time management of the simulation model.

DRMSim simulator is basically based on Java software so that quality of code can have extensibility and re-usability.

Till date, RIP, BGP and NSR routing protocols are being implemented in DRMSim simulator.

The latest version of DRMSim is 1.3.1 which was released on March, 2013.

The following figure shows DRMSim Architecture.

DRMSim has the following features cum properties.

- DRMSim is a discrete event simulator written in Java Language.
- DRMSim has modular design which means new topologies, routing protocols and dynamic models can easily be deployed and simulated in this simulator.
- DRMSim monitors each and every node on the network which means the data structure wise, this simulator is highly efficient.

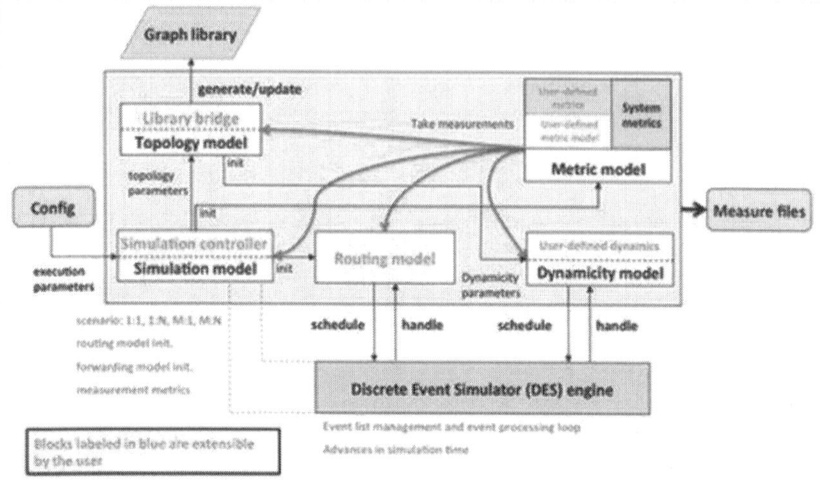

Figure 25. DRMSim Architecture

DRMSim comprise of following components.

- Topology Generators like Brite, Inet, GLP etc.
- Import external topologies. E.g. CAIDA maps.
- Supports high end routing protocols like BGP, DSDV up to 16000 nodes.
- Metric Computation Module- Calculates performance metrics like routing path, routing table size, communication cost etc.

DRMSim is an active simulation project under three year EULER exploratory research project via Grant No: 258307 and currently the progress is going on Geometric Routing, Greedy Routing Models.

16. NetSim [42] [72] [73]

NetSim- Network Simulator is a Stochastic Discrete Event Simulator used for protocol modeling, Network Research and Development and Defense Applications. NetSim provides a unique platform for simulating various types of networks like MANET, WSN, Cognitive Radio, Cellular-GSM, CDAM, Wi-Fi: 802.11 a/b/g/e/n/ac and even more. It plays a crucial role in developing and simulating new application models- Voice,

Video, FTP, HTTP, Email and more. It also features detailed packet animation and information cum packet and event trace. It enables the users to do custom code development and debugging. NetSim provides both Command Level and Graphic Interface to enable users to deploy, simulate and analyze various network scenarios.

The Latest version of NetSim from Boson is version 7.2.

The following figure shows the screenshot of NetSim GUI along with Performance Metric:

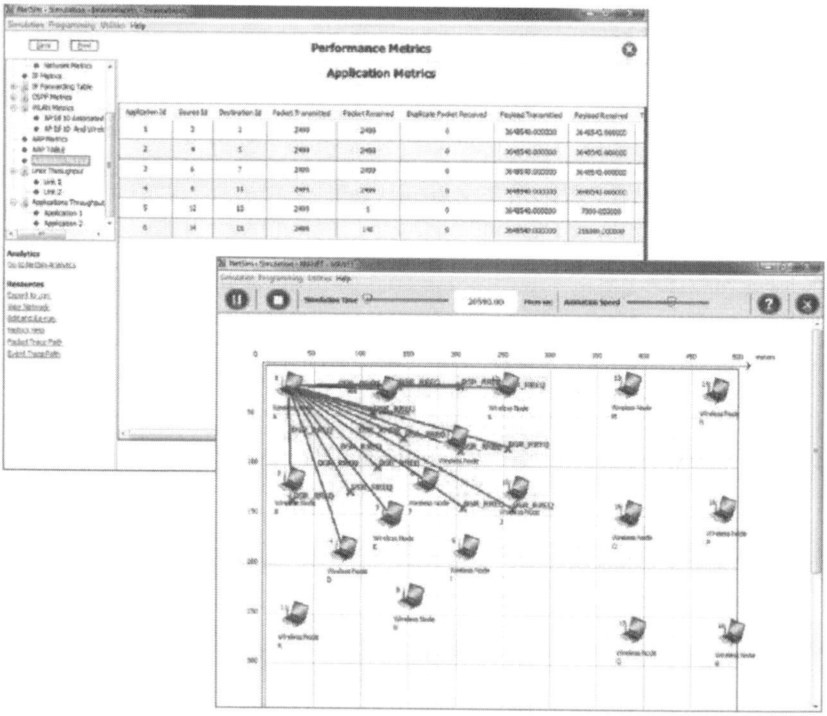

Figure 26. NetSim GUI

NetSim is available as Standard Version which is also known as Academic Version along with Advanced Version.

The following Table 5 demonstrates latest protocols cum functionalities which are available in NetSim v8 for various networks:

Table 5. Protocols Supported by NetSim Simulator

Type of Network	Protocols/Facilities
MANET	DSR, AODV, Mobility via Random Waypoint and Random Walk Models; Codes for various sorts of attacks like Node Failure, Sinkhole, Blackhole etc.
Wireless Sensor Network (WSN)	MAC Layer covering GTS, CCA, Slotted CSMA/CA, Unslotted CSMA/CA, Beacon transmission, LEACH Protocol, PHY Models, Output via Routing Overhead, Power, Lifetime, Packet Delivery Ratio etc.
Cognitive Radio	MAC Layer- DS-MAP, US-MAP, BW Request, Quiet Period, Normal Mode Operation, PHY, Connection Establishment, Others like Spectrum Manager, Sensing Function, Sensing automation and channel model
LTE	Network Later, MAC Layer, PHY Layer, MIMO Model, HARQ Model
CCNA/CCNP Courses	Advanced functionalities- EIGRP, HSRP, VRRP etc.

The following figure highlights the architecture of NetSim 8 Simulator-Simulation Scenario:

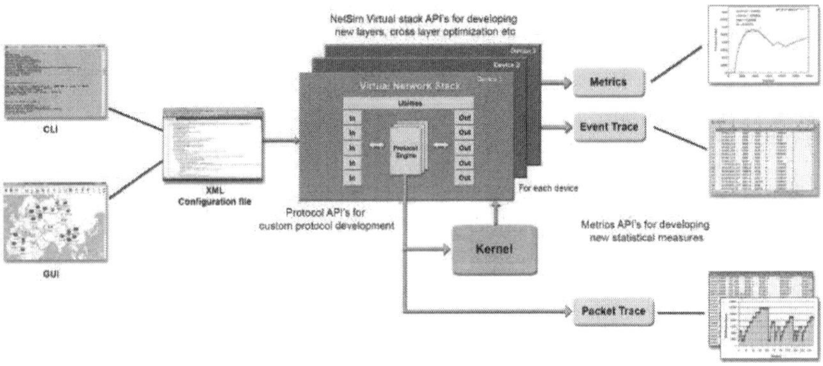

Figure 27. NetSim 8 Simulation Architecture

17. UWSim [4] [74] [75] [76] [77]

UMSim- UnderWater Simulator, an open source tool with prime focus towards marine robotics research and development. UWSim was a joint venture development of RAUVI and TRIDENT research projects. As most of the current available network simulators mainly focus on ground-based sensor systems and Adhoc networks but underwater simulation is usually ignored.

UWSim primarily focus on handling research and simulation scenarios of underwater sensor environments e.g. Low bandwidth, low frequency, high transmission power and limited memory. UWSim is basically designed to handle network scenarios of underwater communication.

UWSim simulator has been implemented in C++ and makes use of OpenSceneGraph (OSG) and osgOcean. OSG is open source 3D graphics application programming interface used by developers in various applications like Virtual Reality, Gaming, Scientific Visualization, Visual Simulation etc. osgOcean is basically used for rendering the ocean surface and special effects.

Windows XP was the prime operating system taken for development of UWSim along with .net Framework 2.0.

The latest version of UWSim is 1.1 and only supports Ubuntu Operating system version 11.04 onwards.

UWSim can only simulate underwater sensor network and no other sensor network can be simulated using this software.

The following figure shows the GUI interface of UWSim Simulator.

18. Visual Sense [78] [4] [27] [28] [71] [79]

Visual Sense- Visual Sense is a modeling cum simulation software for wireless and wireless sensor networks which was built on Ptolemy II. It supports component based modeling, representation and analysis of communication channels, sensors, ad-hoc networking protocols, localization strategies, media access control protocols and even energy consumption in sensor nodes etc. As an extension to Ptolemy it consists of new Java classes and XML files. It supports actor-oriented definition of network nodes, wireless communication channels, physical media such as acoustic channels and wired subsystems. The software architecture consists of set of base classes for defining channels and sensor nodes, library of sub classes which provide channel modes and node models along with visualization framework.

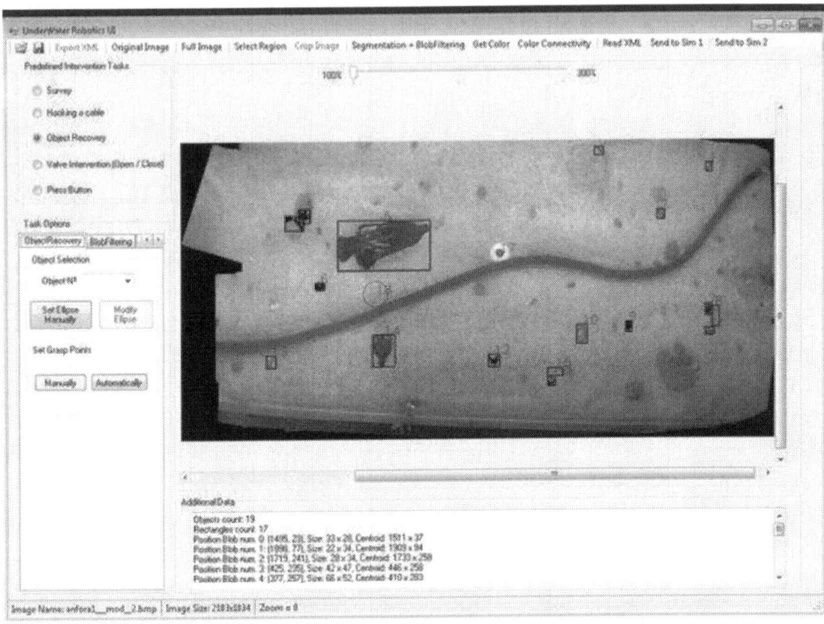

Figure 28. GUI Interface of UWSim

Visual Sense simulator is part of Ptolemy II and the latest version of Visual sense is 8.0.1 which is part of Ptolemy 8.0.1 release and was released on October 28, 2010.

The following figure shows the GUI interface of Visual Sense Simulator:

Visual supports sensor networks, antenna networks, terrain models and battery models. But it doesn't provide any protocols above wireless medium or any sensor or any network other than sound.

19. Viptos [28] [27] [80] [81] [82] [83]

Viptos (Visual Ptolemy and TinyOS) is an integrated graphical development and simulation environment for TinyOS based Wireless Sensor Network.

Viptos is basically an interface/bridge between TinyOS and Ptolemy II and provides a GUI interface for TinyOS developers.

The following figure shows Wireless Communication Simulation in Viptos.

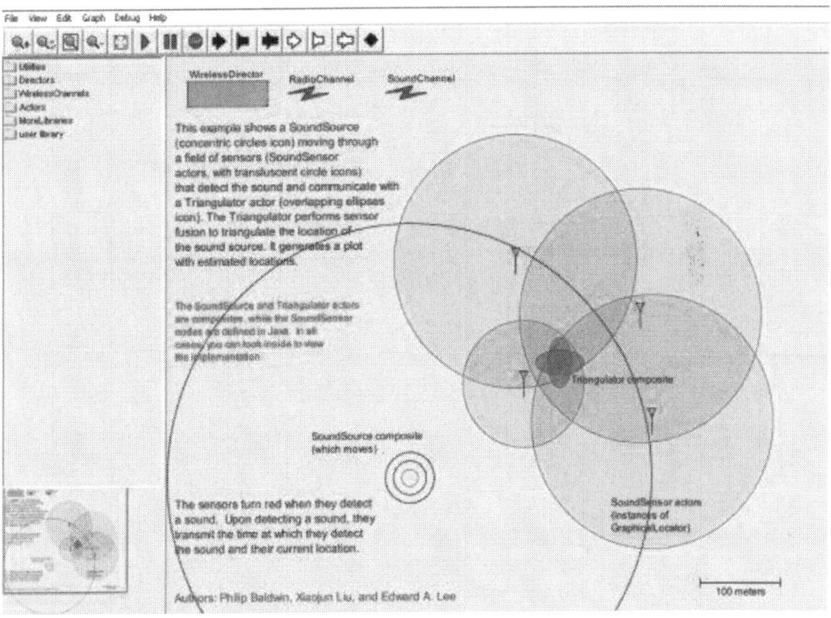

Figure 29. GUI Interface of Visual Sense

With the help of Viptos, the embedded system developers can construct block and arrow diagrams to create TinyOS programs written in nesC, C based programming language. The main advantage of Viptos simulator is that it can transform nesC program into TinyOS supported platform.

The Latest version of Viptos is 1.0.2 which was released in 2006 along with Ptolemy II 6.0.2 with following new features:

- Compatibility with Cygwin
- Multihop routing using TinyOS Surge Protocol
- Java exception handling.

- Improved User Interface.

Viptos has full capability of Visual Sense simulator which is also based on Ptolemy II to model communication channels, networks and non-TinyOS nodes.

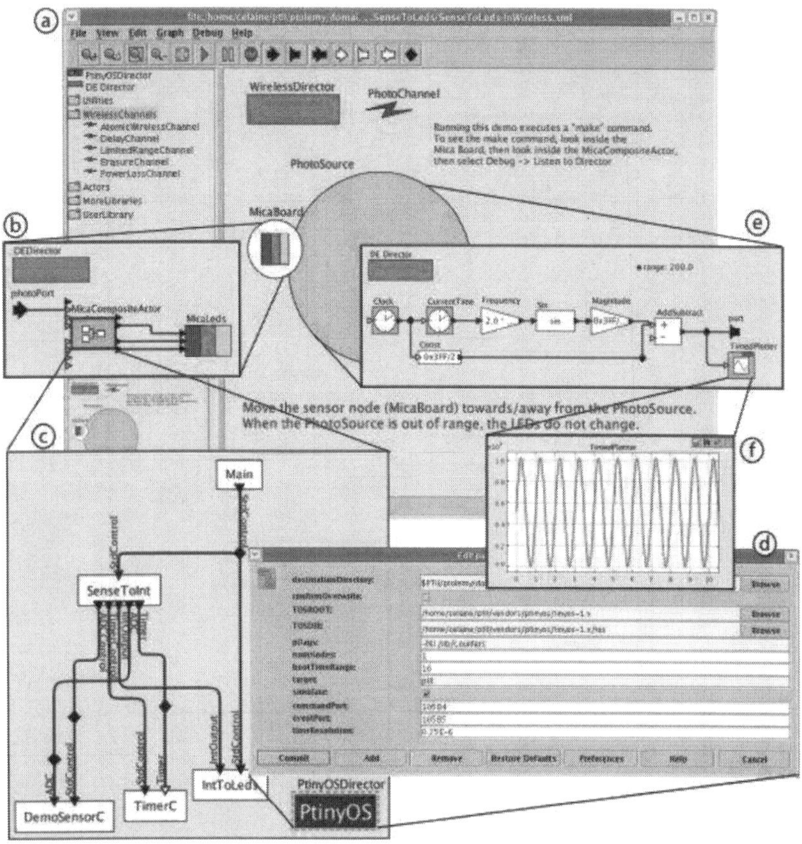

Figure 30. GUI Interface of Viptos-Wireless Simulation

20. PTOLEMY II [6] [84] [71]

Ptolemy II an open source simulation software with actor-oriented design. Actors are defined as software components which execute concurrent and communicate through messages sent via interconnected ports. Ptolemy II is under rigorous development since 1996 and is collection of Java classes and packages. It addresses the modeling, simulation and design of concurrent, real-time, embedded systems.

Ptolemy doesn't support network emulation but is regarded as good simulator having strong features and libraries for simulating both wired and wireless networks.

It comprise of graphical editor Vergil to support editing of syntax. Various specialized simulation tools like HyVisual (Hybrid Systems Modeling), Kepler, Visual Sense, Viptos has been created via Ptolemy.

The current version of Ptolemy II is 10.0.1 and was released on December 17, 2014.

The following figure highlights the GUI based interface of Ptolemy II simulator for simulation of Network Models:

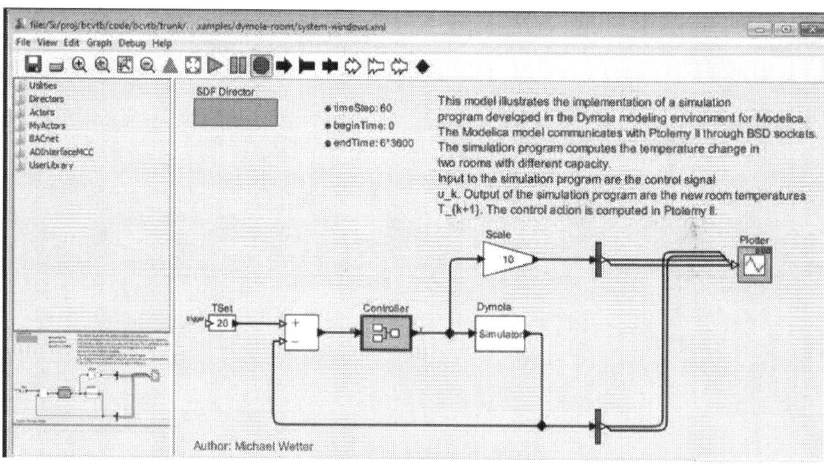

Figure 31. Ptolemy II GUI Interface

The following Table 6 will highlight some of the features of Ptolemy II Simulator.

21. SENS [27] [5] [28] [17] [4] [85] [86] [87]

SENS-Sensor Network Simulator and Emulator is a component based application level wireless sensor network simulator and has feature rich ability of simulation via extensibility and reusability. It has modular and layered architecture and has four main components: (1) Application; (2) Network; (3) Physical; (4) Environment.

Table 6. Features of PTOLEMY II Simulator

Feature Name	Description
Java Language Development	As Ptolemy is based on Java language, so it is platform independent, threaded and network aware software.
Modularity	Comprise of various software's which can be used independently.
Abstract Syntax and Semantics Separation	It defines a software kernel which comprise of various classes which lays the foundation of hierarchical graph.
Heterogeneity	Combination of data flow and non- dataflow is easier.
Mutatable Systems	Design can be modified even when the simulator is running.
Software Architecture	It is designed via taking object modeling and design patterns and test suite is implemented in TCL Blend via TCL/Java Interface.
User Interface	Modular interface using Diva utility
Improved Design	Implementation and coding part is simple to design and interpret.

The following figure demonstrates the components of SENS Simulator:

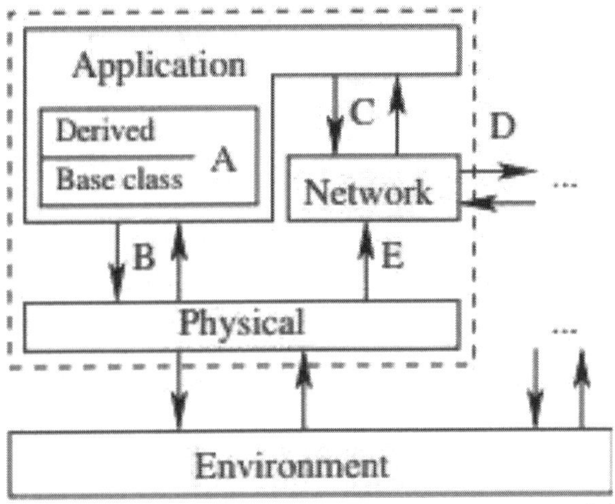

Figure 32. Components of SENS Simulator

SENS comprise of several simulated sensor nodes interacting with environment and each nodes has three main components: Application, Network and Physical. SENS is a customizable sensor network simulator which facilitates users to assemble application-specific environments and such environments are modeling in SENS by various signal propagation characteristics. The same source code which is being deployed on sensors in simulation can be as such used in real sensors which makes this simulator portable because the programming language used is C++.

In light of Wireless Sensor Networks, SENS has novel mechanism for modeling physical environment because of tight integration with real sensors in terms of computation, communication and interaction.

As compared to other WSN simulators, SENS has less ability for customization, so no change can be brought to MAC protocol along with other low level network protocols. SENS does appear to use one of the most sophisticated environmental models and facilitate the use of sensor as well. But only phenomenon which is detected is Sound. But still lots to be done in this simulator and question still arise that whether it is a better simulator or not.

The Latest version can run with GCC 3.3 and was launched on Jan 31, 2005.

The Following Table 7 highlights the network components of SENS Simulator:

Table 7. Network Components of SENS Simulator

Component Name	Component Description
Simple Network	immediate, guaranteed delivery to all neighbours within range
Prob Lossy Network	Probabilistic delivery and delay; delivery probabilities can optionally decrease under heavy traffic.
Collision Lossy Network	calculates collisions at receiving end based on message overlap and relative signal strengths; selectable interval size

22. SHAWN [4] [17] [27] [88] [89] [90]

SHAWN- A Discrete event Open Source Simulator for wireless sensor network is written in C++ language. The basic objective behind the development of SHAWN Simulator to make it fully customizable. One central approach of Shawn is to simulate the effect caused by a phenomenon, not the phenomenon itself. Example: Instead of simulating a complete MAC layer which includes radio propagation model, its effects i.e. Packet Loss and Corruption are only modeled in Shawn.

Shawn provides support for development and testing of high level algorithms cum distributed protocols and has the ability to simulate large scale networks with ability to accommodate a network scenario of more than 100k of nodes.

The following points highlights Shawn Protocol Development Cycle:

- Centralized Implementation: In this phase, the first results and performance are being noted down.

- Simplified Decentralized Protocol: In this phase, proving has to be done that protocol runs in decentralized manner and produce valid results.

- Fully Distributed Protocol: The protocol developed is finalized by seeing the parameters- number of messages, consumption of energy, runtime and resilience to message loss.

The following figure highlights the Architecture of SHAWN Simulator:

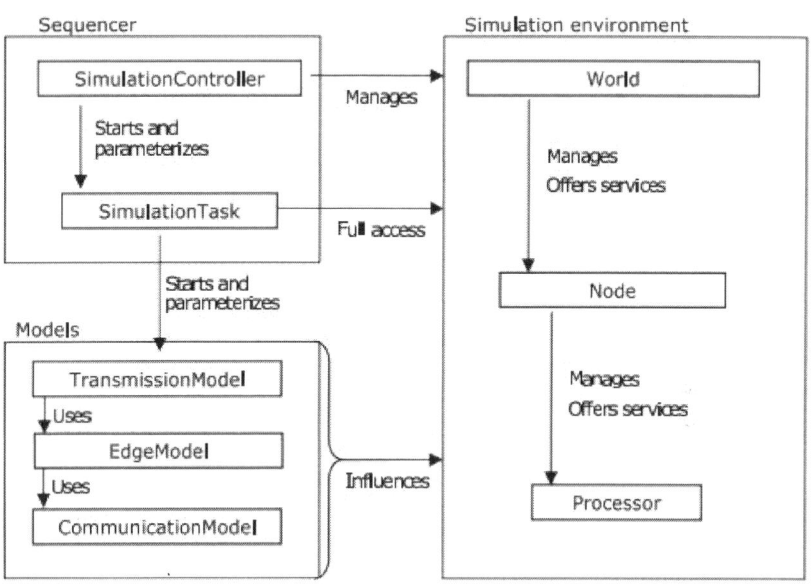

Figure 33. Architecture of Shawn Simulator

The following three models lay strong foundation of Shawn Simulator:

a) Communication Model: It determines if two nodes can communicate by means of exchanging messages.

b) Edge Model: It is responsible for different ways of calculating and maintaining graph representations (connection maps) of the network.

c) Transmission Model: Transmission models are used to decide whether the message will get delayed, corrupted or dropped.

Shawn simulator can be extended to support WISEBED APIs and standards and can read and write WiseML files and provide an iWSN and Session Management implementation to be even connected to federated testbeds, and supports the Wiselib to run an algorithm in simulation before bringing it to a real testbed.

The following figure shows the GUI interface of Satellite Simulation in SHAWN Simulator:

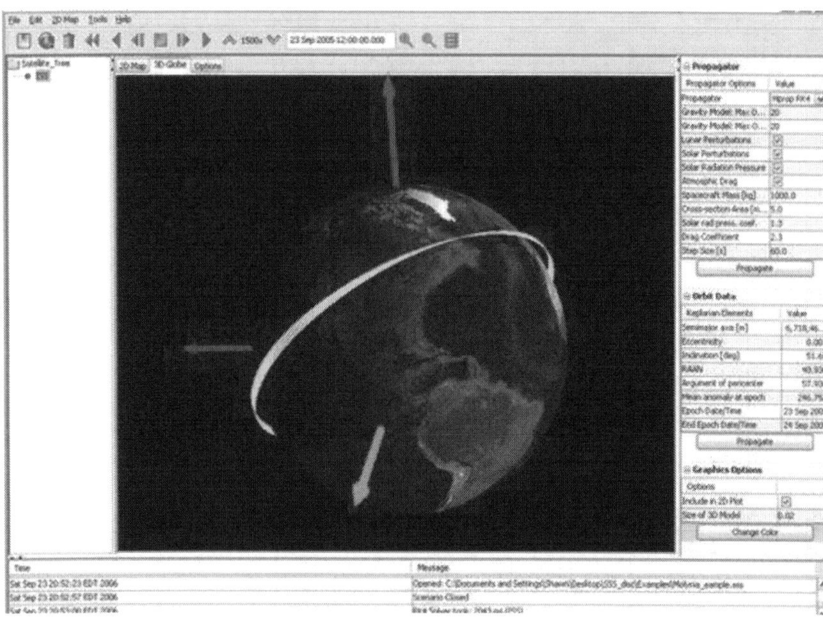

Figure 34. GUI Interface of Satellite Simulation in Shawn Simulator

The Latest version of Shawn in 1.1.6 and is supports both Windows via Cygwin and Linux Operating System. But there is no stable release of this simulator. And in addition to the Shawn Simulator a graphical visualization tool called "Viz" is provided with the simulation package.

23. SIDnet-SWANS [27] [91] [92] [93] [94]

SIDnet-SWANS is a Java-based visualization tool for simulating various application of Wireless Sensor Networks. SIDnet-Swans is based on JiST-SWANS simulator architecture and is used to implement IEEE 802.15.4 MAC-layer protocol. SID-net-SWANS is an efficient simulator to enable end-users to observe the behavior of new algorithms in various different phenomenon's like sudden loss of service both at an individual node as well as collection of nodes in network scenario, unexpected increase/decrease in measurement values, change in network load or increase in network nodes. The simulator also enables users to define various other parameters like temperature, humidity and dynamic movement of nodes in the network.

The following are the features of SIDnet-SWAN:

- Flexible GUI interface for users to interact with Wireless sensor network.
- Real-time data and visual analysis of sensor nodes.
- Explicit control over simulation speed.
- Efficiency towards modeling and management towards energy area of wireless sensor networks.
- Various parameters like Temperature, Humidity and dynamic movements can be monitored because of modular design.

The following figure highlights the SIDnet-SHAWN Simulator architecture.

The components of architecture are listed as follows:

- Sensor Field: It is the area where sensor nodes are placed and sensor field can be setup by the user.
- Utility Views: Utility views are basically user-defined tools such as displaying the energy map and other types of statistical information of the sensor nodes simulating in the environment.

- Simulation Control Interface: It allows the user to control the speed of the simulator.
- Progress Bar: It is used to set the visual feedback regarding any operation of the simulation being performed on the simulator.

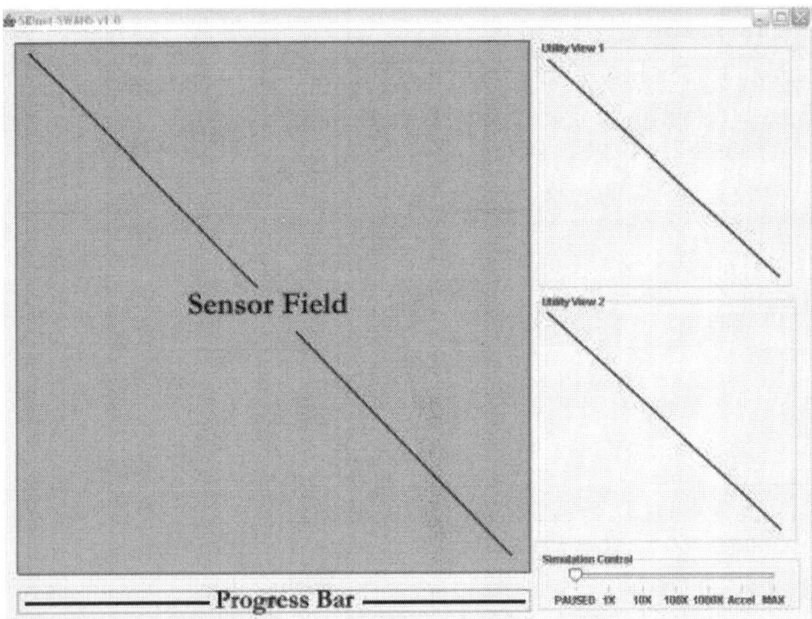

Figure 35. SIDnet-SHAWN Architecture

The following figure highlights the Sensor Network Real-Time Simulation being performed at SIDnet-SWANS Simulator.

The latest version of SIDnet-SWANS is version 1.5.6 which was released in June 2011 and also contains JiST-SWANS version 1.0.6. SIDnet-SWANS simulator project.

24. WSim/Worldsens Simulator/WSNet Simulator [27] [95-98]

WSim/Worldsens-An Event Driven wireless sensor network simulator provides an integrated environment for development cum rapid prototyping of sensor network applications. WSNet simulator models applications, protocols as well as radio medium communication with high amount of efficiency cum accuracy. The objectives behind the development of WSim development are debugging, profiling and performance evaluation for deeply embedded Wireless Sensor Network applications.

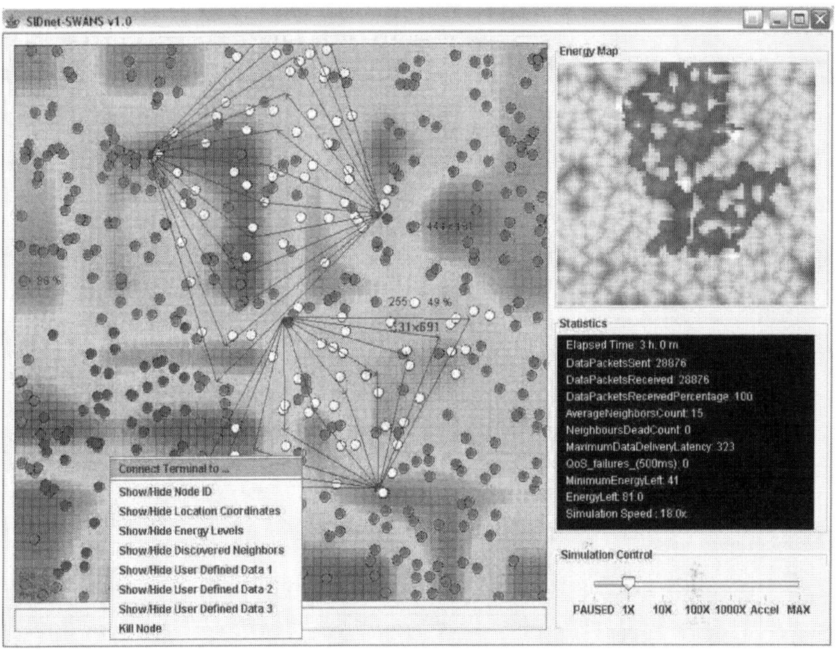

Figure 36. Wireless Sensor Network Simulation being performed at SIDnet-SHAWN Simulator

WSim simulator can be used any sort of embedded operating system like Tiny OS, ContikiOS, FreeRTOS, MANTIS OS etc. and can simulate various real-time sensor nodes like Teas Instruments MSP430, Tiez 430 USB or Olimex Easyweb2.

The following figure highlights the architecture of WSNet Simulator.

The latest stable release of WSNet Simulator is version 9.07 and can be installed across different platforms like Windows, Linux, Solaris and MAC operating systems.

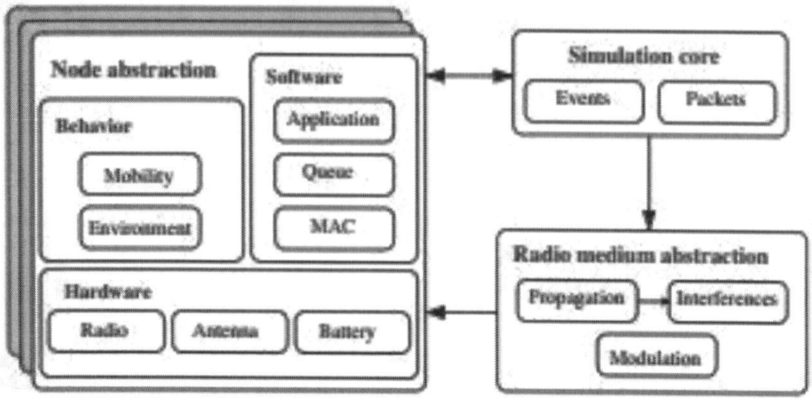

Figure 37. WSNet Simulator Architecture

25. WSN Localization Simulator [27] [99]

Localization is one of the serious issues surrounding proper deployment of sensor nodes in the real world. If the deployment of the sensor nodes is not correct, the results and data will not lay any efficiency. So, how the sensor nodes would be deployed is basically the task of WSN Localization simulator.

The latest version of WSN Localization Simulator is 2.1 [99] and is available open source.

WSN Localization simulator comes bundle with 8 Localization Algorithms: HiRLoc, SeRLoc, Centroid, PTA, ADLA, HADLA, DIL, RAL. Various network parameters can be set like: Number of Random Locators, Number of Static Sensor Nodes, Number of Mobile Sensors, Number of Beacon Nodes and it provides almost all parameters to the researchers to research on the area of localization and has easy and neat and clean GUI for interface and output can be stored in .xls file also.

The following figure highlights GUI interface of WSN Localization Algorithm.

26. NetTopo Simulator [100-105]

NetTopo is regarded as extensible cum integrated framework for simulating and visualizing wireless sensor networks. The basic objective behind design and development of NetTopo Simulator is to analyze various algorithms in WSN. With regard to simulation, users can define large number of parameters over sensor nodes like energy consumption, bandwidth management and NetTopo efficiently takes into consideration large-scale heterogeneous networks. NetTopo is an open source simulator and has been designed in Java language and has more than 80 Java classes and 11000 lines of code.

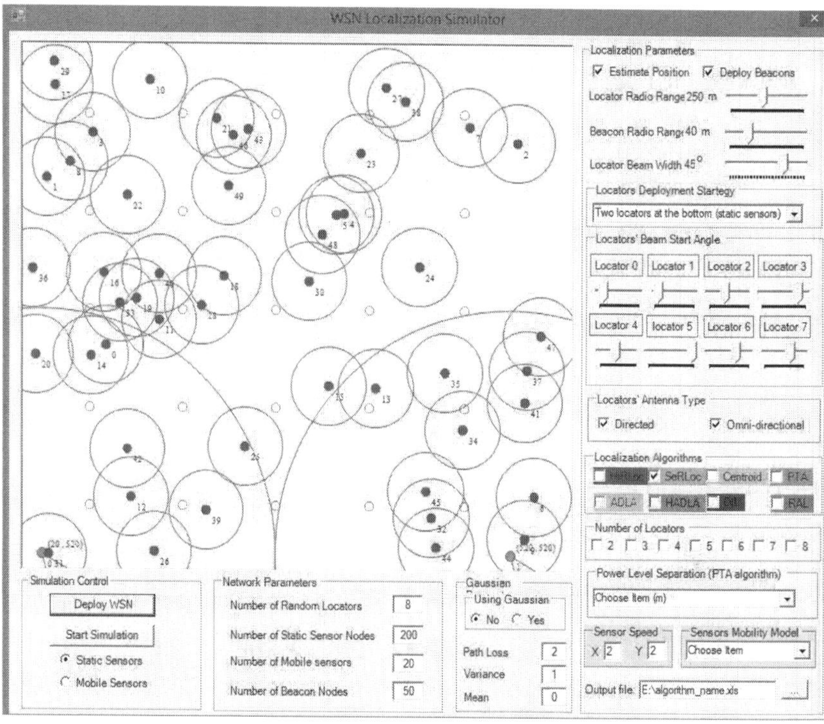

Figure 38. WSN Localization Simulator GUI Interface

Two routing protocols i.e. TPGF and GPSR are already been implemented in this simulator. In addition to simulation, NetTopo has a visualization tool to present the data gathered on actual WSN testbed.

The following figure shows the GUI Interface of NetTopo Simulator:

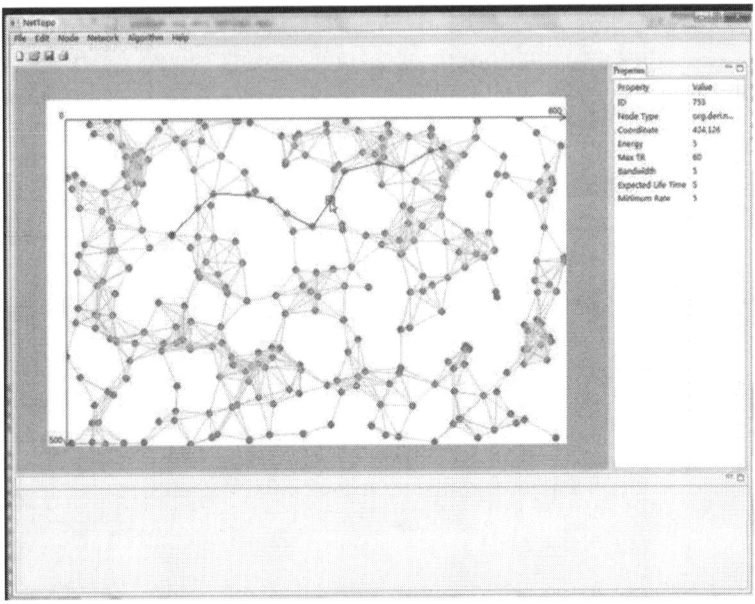

Figure 39. GUI Interface of NetTopo Simulator

The following figure highlights Visualization Tool of NetTopo Simulator:

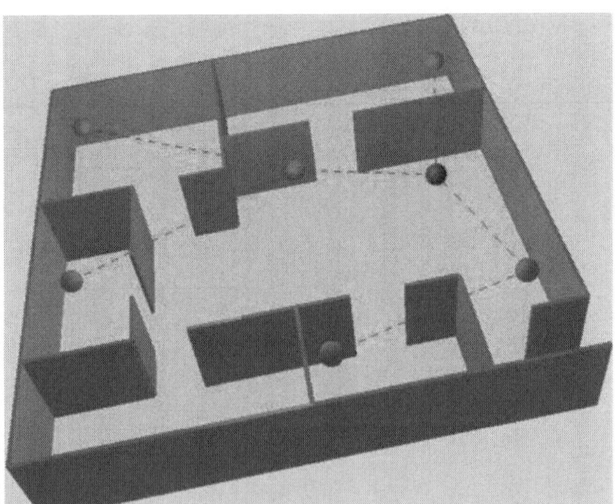

Figure 40. Visualization Tool of NetTopo Simulator

The following figure highlights the Architecture of NetTopo Simulator:

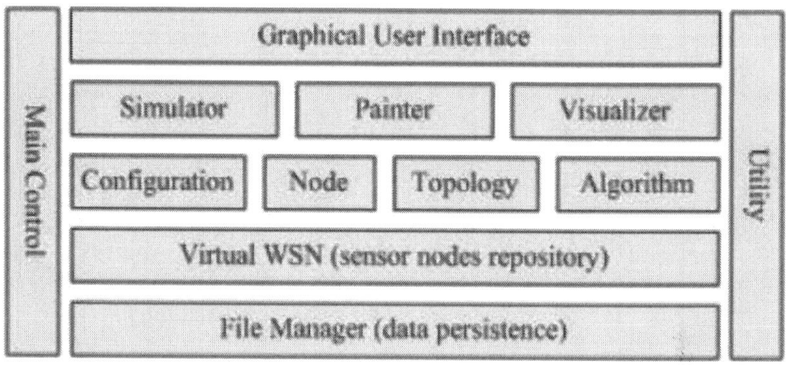

Figure 41. NetTopo Architecture

The following Table 8 explains the various modules/components of NetTopo Architecture:

Main Control and Utility and two main components involved in Architecture of NetTopo Simulator

The following is the list of features of NetTopo Simulator:

- Easy customization of WSN network topology
- Easy customization of properties of sensor nodes from user-defined attributed.
- Extensibility to simulate user-customized algorithms.
- File-based function facilitating users to save and retrieve their simulating process.
- GUI Interface and support for XML file import and export
- 2D and 3D Easy to use Graphical Visualization Tool.

Table 8. Components of NetTopo Simulator

Component Name	Description
Main Control	Acts as coordinator of all the interactions of other components. Works as bridge between input and output interfaces so that simulation operations can be carried out in an efficient manner.
Utility	Performs the basic services: defines application exceptions, format verification, number transforms and dialogue wrappers.
File Manager	It performs various duties like: maintain log of runtime simulation, statistical results recording, and records of all sensor nodes.
Virtual WSN	It does the storage of references of virtual sensor nodes in serialized format so that recovery and reusability tasks can be carried out when necessary.
Node	It is basically a virtual sensor node deployed in the GUI interface for WSN simulation. In NetTopo, various properties of sensor nodes can be configured like temperature, humidity, vibration and pressure etc.
Topology	It is basically the structure of the WSN network which is being designed by user for research and carrying out simulation tasks.
Algorithm	It represents the algorithm to be applied in Sensor network. NetTopo supports various types of algorithms of routing, clustering, scheduling, controlling etc.
GUI	GUI is basically the interface provided to the user to interact with the simulator. It comprise of three main components: (a) Display Canvas- Used for viewing WSN network. (b) Painter- It is an abstract interface for defining various 2D or 3D painting requirements. (c) Visualizer and Simulator: Both required to add sensor nodes on Display Canvas to carry out simulation tasks.

The NetTopo simulator is supported on Windows and Linux operating system.

27. SIDH [106] [5]

SIDH- A Component-based discrete event simulator written in Java is especially designed for wireless sensor network. The aim behind development of SIDH simulator is to simulate thousands of sensor nodes even faster than real-time or any other available WSN simulator. SIDH is easily to reconfigure and adaptable to different levels of simulation detail and accuracy; various communication media; sensors and actuators; environmental conditions and various protocols and applications.

The following Table 9 gives the detailed description of various modules which SIDH simulator is being comprised of:

Table 9. Modules of SIDH Simulator

Module Name	Description
Simulator	It is the foundation module of the simulator. It is in this module various other modules interact with each other.
Event	It is an abstract base class which provides basic functionality to all events. Events tell the time at which an event should fire.
Medium	Medium plays the role of how the sensor nodes will broadcast the signals. Medium lays a strong foundation for the communication model as when the node is activates, what are the changes in the node, node transmission power and other factors like bandwidth and wavelength of medium.
Propagation Model	It defines the signal strength from receiver to transmitter. Various propagation models are used for signal propagation.
Environment	It has somewhat same functionalities as medium module which comprise of physical properties under which sensor operates. Environment also contains physical phenomenon's such as temperature, light, humidity, pressure etc.

Node	It is basically a sensor node which makes up the sensor network. Every node has various components like Processor, transceiver, sensor, actuator, energy source, network protocol and applications.
Transceiver	It models the hardware of sensor node that whether the sensor node is in sleep or active or receive or transmit state.
Protocol	It defines the various rules and regulations of packet movement among the nodes. Protocol can be physical, MAC, Routing or any other protocol.
Application Layer	It is the top layer of the network stack. It interface with other lower layers to implement a WSN application.

28. PROWLER [107-109]

PROWLER- Probabilistic Wireless Sensor Network Simulator- An event driven simulator for wireless sensor network runs under MATLAB and can operate either in deterministic or probabilistic mode. The objective behind development of Prowler simulator was to simulate Berkeley MICA Mote operating on TinyOS and other general systems.

Prowler was basically designed in MATLAB so it provides easy interface to prototype various applications and nice visualizations.

Prowler Simulator provides an accurate Radio Model but only provides a single MAC Protocol which is in Tiny OS.

The current version of Prowler v 1.24 which was tested on Matlab 6.5.

The following figure shows GUI Interface of Prowler Simulator:

An extension to Prowler is JProwler. JProwler is written in Java Language.

29. MATLAB/Simulink [110-113]

Matlab- Matrix Laboratory- A very high performance, varied feature rich computation and visualization tool comprising of thousands of features was built by Mathworks Inc. The Latest version currently available for download or purchase for Commercial purposes is MATLAB 2015A.

The most important feature of MATLAB is its easy programming capability and easy platform for users to develop their own custom functions. MATLAB has various toolboxes like Aerospace, Control System Design, Fuzzy Logic, Symbolic Computations, Statistics, Communication and many others.

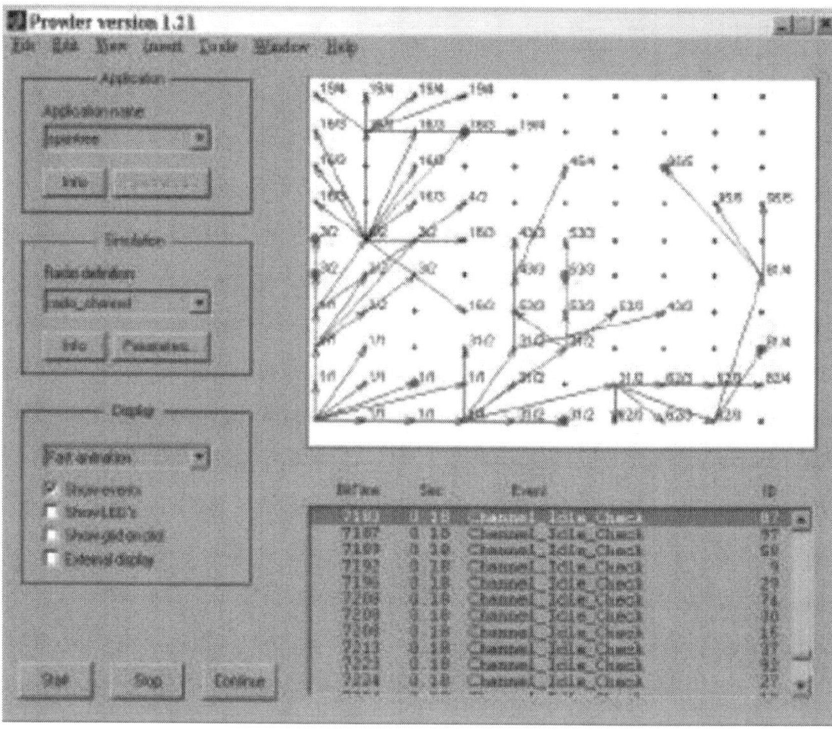

Figure 42. Prowler GUI

The following figure shows basic GUI of MATLAB 2015a software:

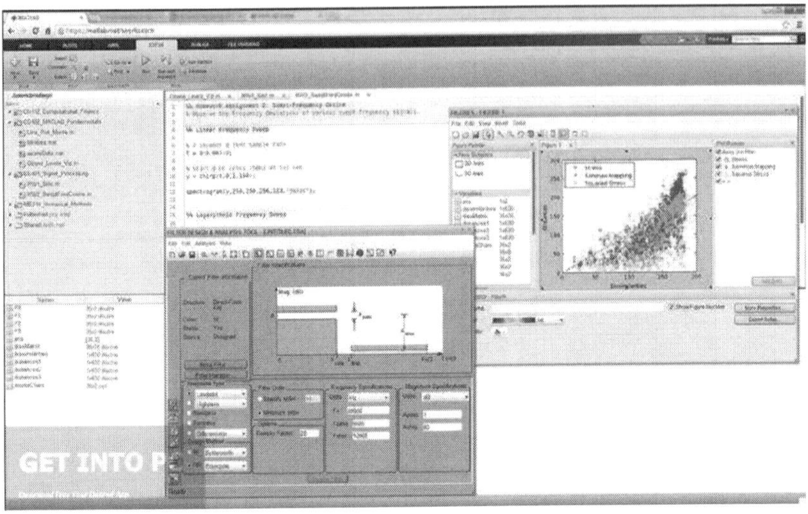

Figure 43. MATLAB 2015a GUI

Matlab is supported by another important software at its backend known as Simulink. Simulink supports both Linear and Non-Linear systems, modelled in continuous time, sampled time or mix of both. Simulink provides GUI for building block diagrams, drag and drop feature to design diagrams and its components.

The following figure shows the GUI interface of Simulink and its library cum components:

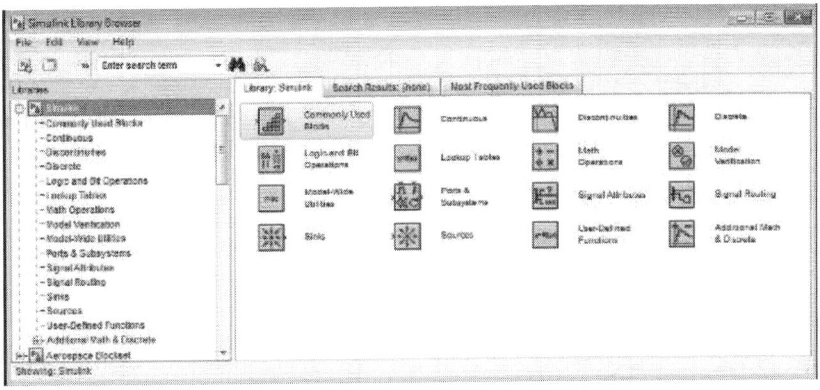

Figure 44. Simulink GUI

With regard to Simulating Wireless Sensor Network, MATLAB and SIMULINK provides communication toolbox set to build a complete WSN model system. Simulation procedure includes building the hardware architecture of transmitting nodes, modeling the communication channel and receiving node architecture. IEEE 802.15.4 was chosen as base standard for physical layer communication and various parameters like SNR (Signal to Noise Ratio, Attenuation and Interference) are considered.

The following figure shows comprehensive detail of WSN Model Architecture developed in Matlab and Simulink:

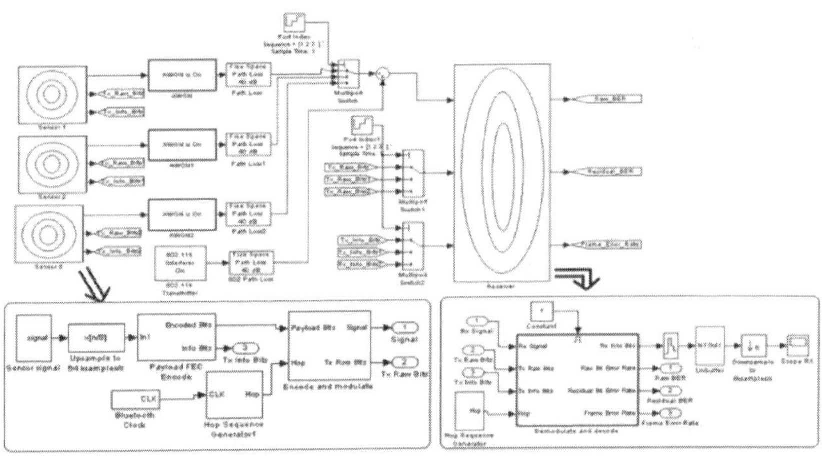

Figure 45. Wireless Sensor Network Model built in Matlab and Simulink

The following Table 10 explains the architecture components of Sensor Node.

30. PiccSIM [115] [116] [117]

PiccSIM- Platform for Integrated Communications and Control design, Simulation, Implementation and Modeling was developed by Wireless Sensor Network Research Group as joint venture between Department of Automation and Systems Technology and Department of Communications and Networking at Aalto University, Espoo, Finland.

Table 10. Architectural Components of Sensor Nodes

Architecture Component	Comprised of:
Transmitter	Operates on Bluetooth at 2.4 GHz and consists of following: • Sensor Signal stage • Up-Sampling to 64ksample/s • Payload FEC encode • Bluetooth clock • Hop Sequence Generator • Encoder and Modulator
Medium	• AWGN Channel • Path Loss • 802.11b interferer • Multiport Switch
Receiver	• Hop Sequence Generator • Demodulation and Decoding • Zero-Order Hold • Un-Buffer • Down-sampling to 8ksamples/s • Scope RX

The idea behind the development of PiccSIM simulator was to make a complete toolset for design, simulation and implementation of Wireless Networked Control Systems (WiNCS) along with Network Controlled System (NCS).

PiccSIM simulator comprise of Simulink (Used for Simulating Dynamic System) and NS-2 (Used for Network Simulation). In addition to this, a GUI interface for modeling and designing the network and control system.

With combination of Simulink and NS-2 PiccSIM can be known as PiccSIM Toolchain which can be used for designing and modeling real wireless nodes.

The following figure highlights PiccSIM Architecture:

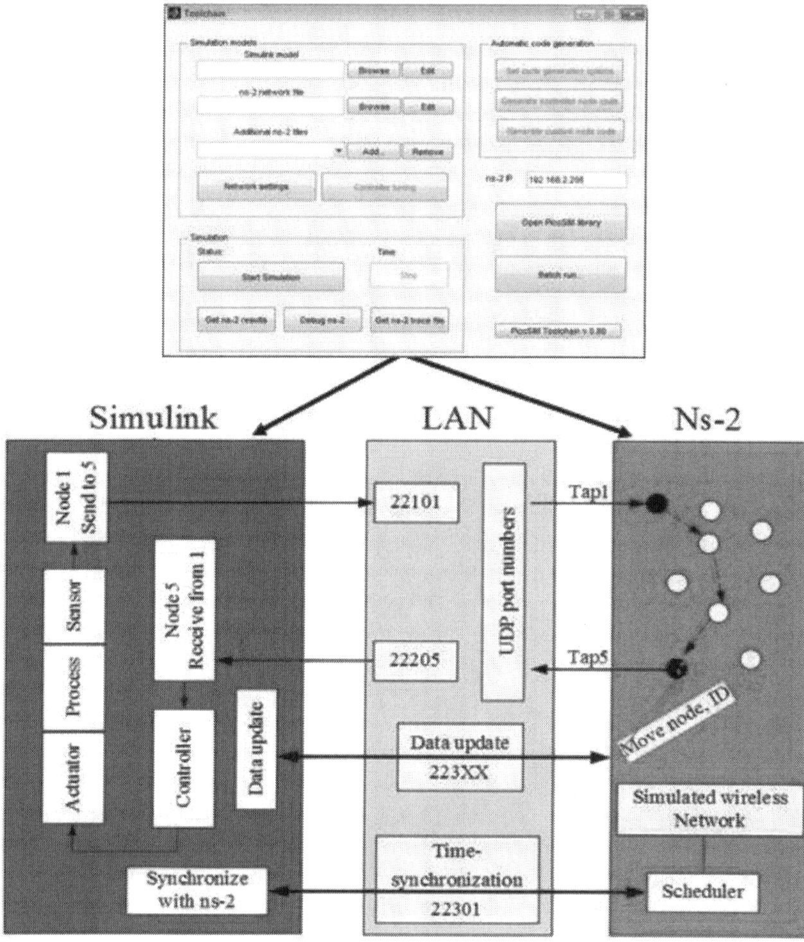

Figure 46. PiccSIM Architecture

PiccSIM makes use of Simulink and NS-2 simulator for NCS simulation. The toolchain uses MATLAB as well as Simulink for controlling the model.

PiccSIM Library comprise of various network node blocks to create wireless communications and Network Setting window provides a GUI interface for running network scripts.

The simulator requires MATLAB and NS-2 to be installed on single machine whether based on Linux or Windows Operating System or might be running on VMware.

The following are the various features of PiccSIM Simulator:

- Wireless Network and Network Controlled Simulation and are combined via Simulink.
- GUI interface for generating network scenario and generating NS-2 TCL scripts.
- Unique data exchange mechanism between Simulink and NS-2
- Complete Automatic code generation of network nodes.
- Remote User Interface which enables simulation without PiccSIM platform.

The latest version of PiccSIM is version 1.16 and now it has been enhanced and a new software of simulation has evolved as True Time-PiccSIM Simulator.

31. LabVIEW [118-121]

LabVIEW (Laboratory Virtual Instrument Engineering Workbench) provides a unique platform and development environment for visual programming language. It is product from National Instruments Inc. LabVIEW these days is most commonly used for data acquisition, instrument control and industrial automation. It supports various operating systems like Windows, Linux, UNIX and even Mac OS X.

LabVIEW provides easy creation of user interfaces known as front panels into development. LabVIEW programs are called Virtual Instruments (Vis). Every VI has three components: Block Diagram, Front panel and Connector Panel.

As LabVIEW comprise of various toolboxes like MATLAB. In LabVIEW 2014 special attention has been given towards integration of Wireless Sensor Network Module. With LabVIEW WSN module, one can add intelligence to NI wireless sensor network nodes via same graphical programming approach to acquire and process data among various sensor nodes. With LabVIEW one can easily incorporate C code in the node for making routing and other all sorts of algorithms.

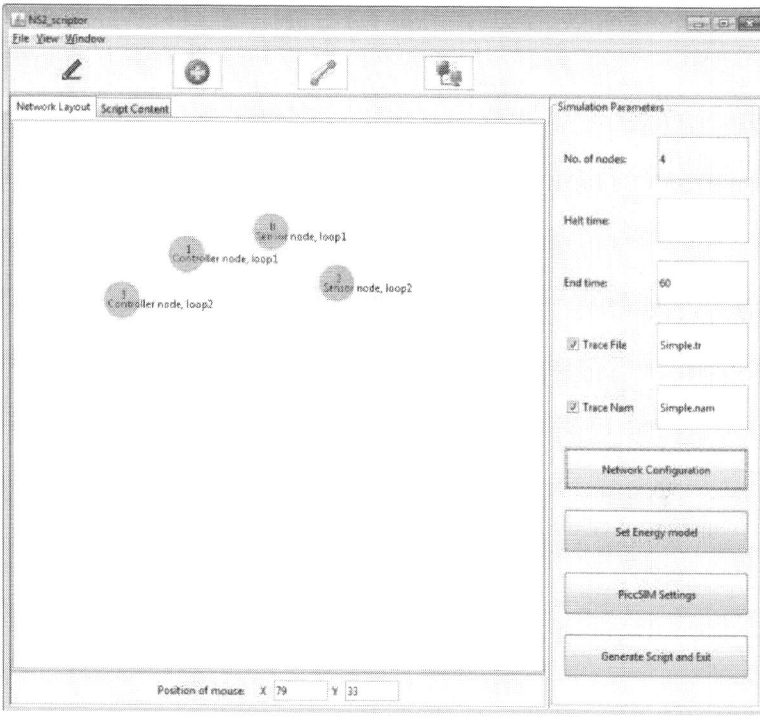

Figure 47. NS-2 Script Generator for PiccSIM Simulator

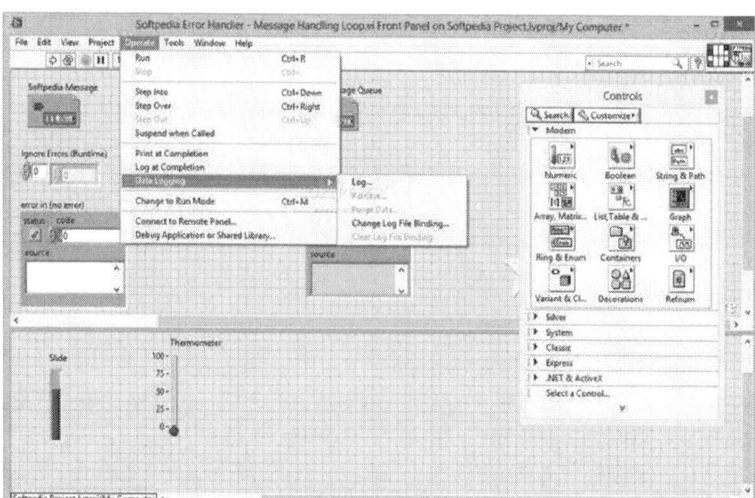

Figure 48. GUI Interface of LabVIEW 2015

The following points highlight some of the features of NI WSN Kit:

- Contains everything to evaluate NI wireless sensor networks and LabVIEW.
- Contains WSN Gateway and programmable ±10 v analog input and thermocouple input nodes.

The following are the main features of WSN module by LabVIEW software:

1) Take advantage of Rapid development with easy to integrate graphical programming.
2) Research done on extension of battery life.
3) Perform custom analysis.
4) Reduce response time with embedded decision making.
5) Reduce costs and increase in performance

The latest version of LabVIEW is LabVIEW 2015/ 2014 SP1.

5. CONCLUSIONS AND FUTURE SCOPE

Simulation is regarded as integral part performed by each and every researcher in various fields to test the research being carried out. In this research paper all the major WSN simulation tools are highlighted which are available till date along with their comparison to guide researchers in choosing the appropriate simulation platform to carry out all sorts of WSN based research.

Considering various parameters taken into consideration for performance comparison of various WSN simulators, NS-2, NS-3 and QualNet Simulator can be considered best simulators to carry out all sort of research in area of Wireless Sensor Networks.

FUTURE SCOPE

In near future, various emulators and testbeds would be evaluated and compared which will give vision to the researcher to test their novel work being done on physical testbeds and on real hardware sensors. A protocol based development would be performed and tested using NS-2/NS-3 simulator.

REFERENCES

1. Akyildiz, I. F., Su, W., Sankarasubramaniam, Y., & Cayirci, E. (2002). Wireless sensor networks: a survey.Computer networks, 38(4), 393-422.

2. Yick, J., Mukherjee, B., & Ghosal, D. (2008). Wireless sensor network survey. Computer networks, 52(12), 2292-2330.

3. Imran, M., Said, A. M., & Hasbullah, H. (2010, June). A survey of simulators, emulators and testbeds for wireless sensor networks. In Information Technology (ITSim), 2010 International Symposium in (Vol. 2, pp. 897-902). IEEE.

4. Sundani, H., Li, H., Devabhaktuni, V. K., Alam, M., & Bhattacharya, P. (2011). Wireless sensor network simulators a survey and comparisons. International Journal of Computer Networks (IJCN), 2(5), 249-265.

5. Singh, C. P., Vyas, O. P., & Tiwari, M. K. (2008, December). A survey of simulation in sensor networks. InComputational Intelligence for Modelling Control & Automation, 2008 International Conference on (pp. 867-872). IEEE.

6. Egea-Lopez, E., Vales-Alonso, J., Martinez-Sala, A. S., Pavon-Marino, P., & García-Haro, J. (2005, July). Simulation tools for wireless sensor networks. InProceedings of the International Symposium on PerformanceEvaluation of Computer and Telecommunication Systems (SPECTS05) (p. 24).

7. Jain, R. (2008). The art of computer systems performance analysis. John Wiley & Sons.

8. Discrete Event Simulation: URL http://en.wikipedia.org/wiki/Discrete_event_simulation#Network_simulators". Accessed on 17 August, 2014.

9. Jevtić, M., Zogović, N., & Dimić, G. (2009, November). Evaluation of wireless sensor network simulators. In Proceedings of the 17th Telecommunications Forum (TELFOR 2009), Belgrade, Serbia (pp. 1303-1306).

10. Mekni, M., & Moulin, B. (2008, August). A survey on sensor webs simulation tools. In Sensor Technologies and Applications, 2008. SENSORCOMM'08. Second International Conference on (pp. 574-579). IEEE.

11. Eriksson, J. (2009). Detailed simulation of heterogeneous wireless sensor networks. Department of Information Technology, Uppsala University.

12. Fall, K., & Varadhan, K. (2005). The ns Manual (formerly ns Notes and Documentation). The VINT project, 47.

13. The Network Simulator-ns2-Available from: http://www.isi.edu/nsnam/ns

14. Issariyakul, Teerawat, and Ekram Hossain. Introduction to network simulator NS2. Springer, 2011.

15. http://www.nsnam.org/docs/release/3.14/tutorial/singlehtml/index.html (Accessed on April 1, 2015)

16. http://www.omnetpp.org/ (Accessed on April 1, 2015)

17. Kellner, A., Behrends, K., & Hogrefe, D. (2010). Simulation environments for wireless sensor networks. no.June, 2.

18. https://castalia.forge.nicta.com.au/index.php/en/ (Accessed on April 2, 2015)

19. Boulis, A. (2009). Castalia User Manual. Online: http://castalia. npc. nicta. com. au/pdfs/Castalia-UserManual. pdf.

20. Varga, A., & Hornig, R. (2008, March). An overview of the OMNeT++ simulation environment. In Proceedingsof the 1st international conference on Simulation tools and techniques for communications,

networks andsystems & workshops (p. 60). ICST (Institute for Computer Sciences, Social-Informatics and Telecommunications Engineering).

21. http://j-sim.cs.uiuc.edu/ (Accessed on April 1, 2015)

22. http://www.physiome.org/jsim/ (Accessed on April 1, 2015)

23. https://sites.google.com/site/jsimofficial/j-sim-tutorial (Accessed on April 1, 2015)

24. Sobeih, A., Chen, W. P., Hou, J. C., Kung, L. C., Li, N., Lim, H., ... & Zhang, H. (2005, April). J-sim: A simulation environment for wireless sensor networks. In Proceedings of the 38th annual Symposium onSimulation (pp. 175-187). IEEE Computer Society.

25. Sobeih, A., Viswanathan, M., Marinov, D., & Hou, J. C. (2007, March). J-Sim: An Integrated Environment for Simulation and Model Checking of Network Protocols. In IPDPS (pp. 1-6).

26. http://www.mannasim.dcc.ufmg.br/ (Accessed on April 1, 2015)

27. Musznicki, B., & Zwierzykowski, P. (2012). Survey of simulators for wireless sensor networks. InternationalJournal of Grid and Distributed Computing, 5(3), 23-50.

28. Steyn, L. P., & Hancke, G. P. (2011, September). A survey of wireless sensor network testbeds. In AFRICON,2011 (pp. 1-6). IEEE.

29. Park, S., Savvides, A., & Srivastava, M. B. (2000, August). SensorSim: A simulation framework for sensor networks. In Proceedings of the 3rd ACM international workshop on Modeling, analysis and simulation ofwireless and mobile systems (pp. 104-111). ACM.

30. http://nesl.ee.ucla.edu/projects/sensorsim/ (Accessed on April 2, 2015)

31. http://www.nrl.navy.mil/itd/ncs/products/sensorsim (Accessed on April 2, 2015)

32. Donward, I. (2004). NRL's sensor network extension to NS-2. Available on line at http://pf. itd. nrl. navy.mil/nrlsensorsim.

33. Jaichandran, R., & Irudhayaraj, A. A. (2010). A Futuristic Analysis on the Application Problems of Wireless Sensor Networks (WSN) with Effective Strategies and Optimal Solutions. International Journal of ComputerApplications, 1(8), 10-16.

34. http://nsl.csie.nctu.edu.tw/nctuns.html (Accessed on April 2, 2015)

35. Wang, S. Y., & Huang, Y. M. (2012). NCTUns distributed network emulator. Internet Journal, 4(2), 61-94.

36. Wang, S. Y., & Huang, Y. M. (2009). NCTUns Tool for Innovative Network Emulations. Computer-Aided Designand Other Computing Research Developments.

37. Wang, S. Y., & Lin, C. C. (2010, May). NCTUns 6.0: a simulator for advanced wireless vehicular network research. In Vehicular Technology Conference (VTC 2010-Spring), 2010 IEEE 71st (pp. 1-2). IEEE.

38. Wang, S. Y., Chou, C. L., Lin, C. C., & Huang, C. H. (2010). The Protocol Developer Manual for the NCTUns 6.0 Network Simulator and Emulator. National Chiao Tung University, Tajwan.

39. Yoon, S., & Kim, Y. B. (2009, September). A design of network simulation environment using ssfnet. InAdvances in System Simulation, 2009. SIMUL'09. First International Conference on (pp. 73-78). IEEE.

40. http://www.ssfnet.org/homePage.html (Accessed on April 2, 2015)

41. Cowie, J. H. (1999). Scalable simulation framework API reference manual. S3 Consortium, March.

42. Kabir, M. H., Islam, S., Hossain, M. J., & Hossain, S. Detail Comparison of Network Simulators.

43. Zeng, X., Bagrodia, R., & Gerla, M. (1998, May). GloMoSim: a library for parallel simulation of large-scale wireless networks. In Parallel and Distributed Simulation, 1998. PADS 98. Proceedings. Twelfth Workshop on(pp. 154-161). IEEE.

44. Bajaj, L., Takai, M., Ahuja, R., Tang, K., Bagrodia, R., & Gerla, M. (1999). Glomosim: A scalable network simulation environment.

UCLA Computer Science Department Technical Report, 990027, 213.

45. GAO, Z. G., ZHAO, Y. L., LI, X., CAI, S. B., & WANG, C. S. (2006). Analysis of GloMoSim Wireless Network Simulator. Journal of System Simulation, S2.

46. Bagrodia, R., Meyer, R., Takai, M., Chen, Y. A., Zeng, X., Martin, J., & Song, H. Y. (1998). Parsec: A parallel simulation environment for complex systems.Computer, 31(10), 77-85.

47. http://web.scalable-networks.com/content/qualnet (Accessed on April 3, 2015)

48. http://web.scalable-networks.com/sites/default/files/PR131023-SCALABLE-Releases-QualNet-and-EXata-Upgrades.pdf (Accessed on April 3, 2015)

49. Vasu, B., Varshney, M., Rengaswamy, R., Marina, M., Dixit, A., Aghera, P., ... & Bagrodia, R. (2005, November). Squalnet: a scalable simulation framework for sensor networks. In Proceedings of the 3rdinternational conference on Embedded networked sensor systems (pp. 322-322). ACM.

50. Varshney, M., Xu, D., Srivastava, M., & Bagrodia, R. (2007, April). sQualNet: A scalable simulation and emulation environment for sensor networks. In Proceedings of the International Conference on InformationProcessing in Sensor Networks, New York, NY, USA. Citeseer (p. 24).

51. Varshney, M., Bagrodia, R., Dixit, A., Vasu, B., Aghera, P., & Srivastava, M. sQualNet version 1.0 beta: Tutorial and Programmers Manual.

52. http://nesl.ee.ucla.edu/projects/squalnet/ (Accessed on April 3, 2015)

53. Sethi, A. S., & Hnatyshin, V. Y. (2012). The Practical OPNET User Guide for Computer Network Simulation. CRC Press.

54. Lu, Z., & Yang, H. (2012). Unlocking the power of OPNET modeler. Cambridge University Press.

55. García-Hernando, A. B., Martínez-Ortega, J. F., López-Navarro, J. M., Prayati, A., & Redondo-López, L. (2008). Software Technologies in WSNs (pp. 1-49). Springer London.

56. Chang, X. (1999, December). Network simulations with OPNET. In Proceedings of the 31st conference onWinter simulation: Simulation---a bridge to the future-Volume 1 (pp. 307-314). ACM.

57. Katzela, I. (1998). Modeling and simulating communication networks: a hands-on approach using OPNET. Prentice Hall PTR.

58. http://www.opnet.com/university_program/itguru_academic_edition/ (Accessed on April 4, 2015)

59. http://www.riverbed.com/products/performance-management-control/opnet.html (Accessed on April 4, 2015)

60. https://support.riverbed.com/bin/support/static//doc/opnet/OnlineHTML/Modeler/modeler_17.5_PL6/ wwhelp/wwhimpl/js/html/ wwhelp.htm (Accessed on April 4, 2015)

61. Chen, G., Branch, J., Pflug, M., Zhu, L., & Szymanski, B. (2005). SENSE: a wireless sensor network simulator. In Advances in pervasive computing and networking (pp. 249-267). Springer US.

62. Szymanski, B. K., & Chen, G. G. (2007). Sensor network component based simulator. Handbook of DynamicSystem Modeling, 35, 1-16.

63. http://www.ita.cs.rpi.edu/ (Accessed on April 4, 2015)

64. Rosa, P. M., Neves, P. A., Vaidya, B., & Rodrigues, J. J. (2009, September). G-Sense-A Graphical Interface for SENSE Simulator. In Advances in System Simulation, 2009. SIMUL'09. First International Conference on(pp. 88-93). IEEE.

65. http://netgna.it.ubi.pt/Downloads/G-SENSE.php (Accessed on April 4, 2015)

66. http://drmsim.gforge.inria.fr/index.php?page=intro (Accessed on April 4, 2015)

67. Hogie, L., Tahiri, I., Papadimitriou, D., & Majorczyk, F. (2010). DRMSim: a network simulator for the investigation of routing schemes User manual.

68. Lancin, A., & Papadimitriou, D. (2013). DRMSim: A Routing-Model Simulator for Large-Scale Networks. ERCIMNews, 94, 31-32.

69. Hogie, L., Papadimitriou, D., Tahiri, I., & Majorczyk, F. (2010, May). Simulating routing schemes on large-scale topologies. In Principles of Advanced and Distributed Simulation (PADS), 2010 IEEE Workshop on (pp. 1-10). IEEE.

70. Chéour, R., Jmal, M. W., Lay-Ekuakille, A., Derbel, F., Kanoun, O., & Abid, M. (2013, October). Choice of efficient simulator tool for wireless sensor networks. In Measurements and Networking Proceedings (M&N),2013 IEEE International Workshop on (pp. 210-213). IEEE.

71. Redondo-López, L., Prayati, A., López-Navarro, J. M., Martínez-Ortega, J. F., & García-Hernando, A. B. (2008). Problem Solving for Wireless Sensor Networks. Problem Solving for Wireless Sensor Networks:,Computer Communications and Networks. ISBN 978-1-84800-203-6. Springer-Verlag London, 2008, 1.

72. NetSim 8 User Manual: Available at: http://www.boson.com/files/support/NetSim-8-User-Manual.pdf (Accessed on April 4, 2015)

73. http://www.tetcos.com/netsim_gen.html (Accessed on April 4, 2015)

74. Dhurandher, S. K., Misra, S., Obaidat, M. S., & Khairwal, S. (2008). UWSim: A simulator for underwater sensor networks. Simulation, 84(7), 327-338.

75. Prats, M., Pérez, J., Fernández, J. J., & Sanz, P. J. (2012, October). An open source tool for simulation and supervision of underwater intervention missions. In Intelligent Robots and Systems (IROS), 2012 IEEE/RSJInternational Conference on (pp. 2577-2582). IEEE.

76. Dhviya, V. P., & Arthi, R. Analysis of Simulation Tools for Underwater Wireless Sensor Networks.

77. http://www.irs.uji.es/uwsim/ (Accessed on April 4, 2015)

78. Kohli, S., Lee, E. A., Liu, X., Zhao, Y., Ee, C. T., Brooks, C., ... & Zhou, R. (2005). Visualsense: Visual modeling for wireless and sensor network systems.

79. http://ptolemy.eecs.berkeley.edu/visualsense/ (Accessed on April 4, 2015)

80. Cheong, E., Lee, E. A., & Zhao, Y. (2005, November). Viptos: a graphical development and simulation environment for tinyos-based wireless sensor networks. In SenSys (Vol. 5, pp. 302-302).

81. Teng, G., Zheng, K., & Dong, W. (2008, October). A survey of available tools for developing wireless sensor networks. In Systems and Networks Communications, 2008. ICSNC'08. 3rd International Conference on (pp. 139-144). IEEE.

82. http://ptolemy.eecs.berkeley.edu/viptos/ (Accessed on April 4, 2015)

83. Cheong, E., Lee, E. A., & Zhao, Y. (2006). Joint modeling and design of wireless networks and sensor node software. EECS Department, University of California, Berkeley, UCB/EECS-2006-150.

84. http://ptolemy.eecs.berkeley.edu/ptolemyII/ (Accessed on April 4, 2015)

85. Sundresh, S., Kim, W., & Agha, G. (2004, April). SENS: A sensor, environment and network simulator. InProceedings of the 37th annual symposium on Simulation (p. 221). IEEE Computer Society.

86. http://osl.cs.illinois.edu/sens (Accessed on April 4, 2015)

87. Peterson, J. A. (2005). SENS: A simple, extensible sensor network simulator for exploring approximate aggregation techniques. Master's thesis, Boston University.

88. Kröller, A., Pfisterer, D., Buschmann, C., Fekete, S. P., & Fischer, S. (2005). Shawn: A new approach to simulating wireless sensor networks. arXiv preprint cs/0502003.

89. http://www.wisebed.eu/#appdev_shawn (Accessed on April 4, 2015)

90. Fekete, S. P., Kröller, A., Fischer, S., & Pfisterer, D. (2007, June). Shawn: The fast, highly customizable sensor network simulator. In Proceedings of the Fourth International Conference on Networked SensingSystems (INSS 2007)(Vol. 32).

91. http://www.ece.northwestern.edu/~ocg474/SIDnet.html (Accessed on April 4, 2015)

92. Ghica, O. C., Trajcevski, G., Scheuermann, P., Bischof, Z., & Valtchanov, N. (2008, November). Sidnet-swans: A simulator and integrated development platform for sensor networks applications. In Proceedings ofthe 6th ACM conference on Embedded network sensor systems (pp. 385-386). ACM.

93. Pathan, A. S. K., Monowar, M. M., & Khan, S. (Eds.). (2014). Simulation Technologies in Networking andCommunications: Selecting the Best Tool for the Test. CRC Press.

94. Ghica, O. C. (2010). SIDnet-SWANS manual.

95. http://perso.ens-lyon.fr/eric.fleury/Upload/wsim-docbook/ tools. Html#id4743825 (Accessed on April 5, 2015)

96. http://wsim.gforge.inria.fr/overview.html (Accessed on April 5, 2015)

97. Fraboulet, A., Chelius, G., & Fleury, E. (2007, April). Worldsens: development and prototyping tools for application specific wireless sensors networks. In Information Processing in Sensor Networks, 2007. IPSN2007. 6th International Symposium on (pp. 176-185). IEEE.

98. Chelius, G., Fraboulet, A., & Fleury, E. (2007, March). Worldsens: a fast and accurate development framework for sensor network applications. In Proceedings of the 2007 ACM symposium on Applied computing(pp. 222-226). ACM.

99. http://www.codeproject.com/Articles/606364/Wireless-Sensor-Network-Localization-Simulator-v (Accessed on April 5, 2015)

100. Shu, L., Wu, C., Zhang, Y., Chen, J., Wang, L., & Hauswirth, M. (2008). NetTopo: beyond simulator and visualizer for wireless sensor networks. ACM SIGBED Review, 5(3), 2.

101. Shu, L., Hauswirth, M., Chao, H. C., Chen, M., & Zhang, Y. (2011). NetTopo: A framework of simulation and visualization for wireless sensor networks. Ad Hoc Networks, 9(5), 799-820.

102. Shu, L., Wu, C., & Hauswirth, M. NetTopo: a framework of simulation and visualization for wireless sensornetworks, Digital Enterprise Research Institute (DERI), National University of Ireland. Galway, Technical Report, October 30, 2008. Available at: http://library. deri. ie/resource/jbug5jlj.

103. Shu, L., Hauswirth, M., Wang, L., Yuan, Z., & Chen, Y. (2009, September). Visualizing simulation and testbed of wireless sensor networks with NetTopo. In Proceedings of the 4th ACM international workshop onExperimental evaluation and characterization (pp. 99-100). ACM.

104. Parbat, B., Dwivedi, A. K., & Vyas, O. P. (2010). Data visualization tools for WSNs: a glimpse. InternationalJournal of Computer Applications, 2(1), 14-20.

105. https://sites.google.com/site/leishuderi/nettopo (Accessed on April 5, 2015)

106. Carley, T. W. (2005). Sidh: A wireless sensor network simulator.

107. Simon, G., Volgyesi, P., Maróti, M., & Lédeczi, Á. (2003, March). Simulation-based optimization of communication protocols for large-scale wireless sensor networks. In IEEE aerospace conference (Vol. 3).

108. Simon, G. (2003). Prowler: Probabilistic wireless network simulator. Institute for Software IntegratedSystems, Nashville.

109. http://www.isis.vanderbilt.edu/projects/nest/prowler/ (Accessed on April 5, 2015)

110. Simek, M., Moravek, P., & sa Silva, J. (2011). Wireless Sensor Networking in Matlab: Step-by-Step. Inproceedings of International Conference on Teleinformatics (pp. 185-190).

111. Ali, Q. I. (2012). Simulation Framework of Wireless Sensor Network (WSN) Using MATLAB/SIMULINK Software. Edited by Vasilios N. Katsikis, 263.

112. Ali, Q. I., Abdulmaowjod, A., & Mohammed, H. M. (2010, November). Simulation & performance study of wireless sensor network (WSN) using MATLAB. In Energy, Power and Control (EPC-IQ), 2010 1st InternationalConference on (pp. 307-314). IEEE.

113. Matlab website: www.mathworks.com

114. http://wislab.cz/our-work/wireless-sensor-network-simulation-tutorial-for-matlab (Accessed on April 5, 2015)

115. Kohtamaki, T., Pohjola, M., Brand, J., & Eriksson, L. M. (2009, March). Piccsim toolchain-design, simulation and automatic implementation of wireless networked control systems. In Networking, Sensing and Control,2009. ICNSC'09. International Conference on (pp. 49-54). IEEE.

116. http://wsn.aalto.fi/en/tools/piccsim/ (Accessed on April 5, 2015)

117. PiccSim Manual: Available athttp://wsn.aalto.fi/en/tools/piccsim/piccsim_manual_1.1.pdf (Accessed on April 15, 2015)

118. Labview Website: www.ni.com/labview (Accessed on April 5, 2015)

119. http://www.ni.com/white-paper/8981/en/ (Accessed on April 5, 2015)

120. http://www.ni.com/labview/wsn/ (Accessed on April 5, 2015)

121. http://en.wikipedia.org/wiki/LabVIEW (Accessed on April 6, 2015)

122. Nayyar, A., & Gupta, A. (2014). A Comprehensive Review of Cluster-Based Energy Efficient Routing Protocols in Wireless Sensor Networks. IJRCCT, 3(1), 104-110.

123. Nayyar, A., & Sharma, S. (2014). A Survey on Coverage and Connectivity Issues Surrounding Wireless Sensor Network. IJRCCT, 3(1), 111-118.

124. Nayyar, A. (2012, January). Simulation Based Evaluation of Reactive Routing Protocol for MANET. InAdvanced Computing & Communication Technologies (ACCT), 2012 Second International Conference on (pp. 561-568). IEEE.

125. Nayyar, A., & Singh, R. (2014). A Comprehensive Review of Ant Colony Optimization (ACO) based Energy-Efficient Routing Protocols for Wireless Sensor Networks. International Journal of Wireless Networks andBroadband Technologies (IJWNBT), 3(3), 33-55

9

EFFICIENT SPECTRUM SCHEDULING AND POWER MANAGEMENT FOR OPPORTUNISTIC USERS

Raouia Masmoudi, E. Veronica Belmega and Inbar Fijalkow

[1]ETIS/ ENSEA, University Cergy Pontoise, CNRS, F-95000 Cergy Pontoise, France.

[2]LETI, ENIS and University of Sfax, Sfax, Tunisia.

[3]INRIA Grenoble Rhône-Alpes, Grenoble, France.

ABSTRACT

In this paper, we study the centralized spectrum access and power management for several opportunistic users, secondary users (SUs), without hurting the primary users (PUs). The radio resource manager's objective is to minimize the overall power consumption of the opportunistic system over several orthogonal frequency bands under constraints on the minimum quality of service (QoS) and maximum peak and average interference to the PUs. Given the opposing nature of these constraints, we first study the problem of feasibility, and we provide sufficient conditions and necessary conditions for the existence of a solution. The main challenge lies in the non-convexity of this problem because of the discrete spectrum scheduling: one band can be allocated to at most one SU to avoid interference impairments. To overcome this issue, we use a Lagrangian relaxation technique, and we prove that the discrete solutions of the relaxed problem are the solutions to the initial problem. We propose a projected sub-gradient algorithm to compute the solution, when it exists. Assuming that the channels are drawn randomly from a continuous distribution, this algorithm converges to the optimal solution. We also study a specific symmetric system for which we provide the analytical solution. Our numerical results compare the energy-efficiency of the proposed algorithm with other spectrum allocation solutions and show the optimality of our approach.

KEYWORDS

Cognitive radio systems; Spectrum scheduling; Power allocation; Lagrangian relaxation; Projected sub-gradient algorithm

1. INTRODUCTION

Most frequency bands in the radio spectrum have already been licensed, and it is difficult to find vacant bands for wireless communication systems. At the same time, the most allocated spectrum is under-utilized [1]. Cognitive radio (CR) systems, as explained in [2] and references therein, propose to better utilize the spectrum by allowing an opportunistic access to it. A hierarchy between users is imposed, in which secondary users (SUs) are allowed, by the spectrum manager, to communicate either in the vacant bands left by the licensed users, called primary users (PUs), or the non-vacant bands under the condition that the created interference (at the primary receivers) is kept below some predefined thresholds [3]. The radio resource manager uses channel state information (CSI) to coordinate the access to the wireless radio spectrum. When performed in a centralized way, this management is often referred to as coordinated multi-point (CoMP) radio resource management [4]. In the CR paradigm, CSI is provided by spectrum sensing at different remote locations and/or by backhaul information feedback from the spectrum manager to improve the spectrum usage. Carrier aggregation and multi-carrier communications have been suggested as promising candidates for both the CR and CoMP systems thanks to their flexible usage of the spectrum [5].

In ad-hoc and sensor networks or even in future 5G, a major bottleneck is the power consumption efficiency caused by limited battery-life device systems and operating costs [6, 7]. In this work, we investigate a centralized power minimization problem with quality of service (QoS) requirements for the secondary users imposed by the spectrum manager. In such a centralized setting, the spectrum manager can be more effective when it operates opportunistic users scheduling in addition to the sole power allocation, as mentioned in [8] and [9]. Very few existing works consider both bandwidth scheduling and power allocation jointly in the CR context. In particular, [10] provides a heuristic algorithm for the users scheduling.

In this paper, we consider a joint discrete scheduling and power allocation problem that aims at a minimal power consumption under QoS and interference power constraints in a centralized CR system. Such a problem introduces two major challenges. First, the minimum QoS and maximum interference constraints may not be simultaneously satisfied. Several works in the wireless communications literature [11, 12] have proposed a classical water-filling procedure to solve rate-driven or power-driven resource allocation problems in classical interference or multiple access channels. In our study, the presence of the PUs imposes additional peak and total interference constraints aside the classical minimum QoS constraint for the SU communication. These additional constraints are the main reason why classical numerical approaches are not suitable. In order to tackle the problem of feasibility, we study necessary conditions and sufficient conditions on the CSI for the existence of a feasible allocation point. The second difficulty raised by the problem under investigation is the discrete nature of channel assignments in the scheduling policy. This policy makes the problem a non-convex optimization one. Inspired by the approach in [13], we use a Lagrangian relaxation and a dual approach to obtain a solvable convex optimization problem. We then study the Karush-Kuhn-Tucker (KKT) conditions of the relaxed problem and show that the solutions meeting these conditions are actually the solutions of the initial non-convex problem. To solve the relaxed problem numerically, we propose a projected sub-gradient algorithm [14] when the problem is feasible.

1.1 Related works

Power allocation problems have been the subject of several studies in non-CR systems from a rate maximization point of view and via centralized [15] or decentralized (using non-cooperative games) [16, 17, 18] approaches. Also, power allocation problems without spectrum allocation in non-CR networks have been studied from an energy-efficiency point of view in centralized [19, 20] and decentralized systems [21, 22]. In this paper, we consider a centralized radio management, to make the spectrum manager more effective, it operates opportunistic user's scheduling in addition to the sole power allocation, as mentioned in [8] and [9]. Such joint resource allocation problems have been the subject of several studies in non-cognitive radio settings such as in code division multiple access (CDMA) systems [23] and in downlink and uplink orthogonal frequency division multiplexing (OFDM) systems [24] and [13], respectively. In [23], the scheduling and resource allocation problem for the downlink in a CDMA-based wireless network is considered. The problem is to select a subset of the users for transmission and, for each of these users, to choose the optimal modulation, coding scheme and transmit power allocation policy. In [24], the authors consider the scheduling and resource allocation for the downlink of a cellular OFDM system. An optimal algorithm is proposed assuming that multiple users can time-share each tone and several low complexity heuristics are that enforce integer tone allocations. Among the works on OFDM [13,24], the closest to our work is [13], in which a dynamic scheduling and power allocation algorithm was proposed to compute the policies of the multiple non-interfering users that maximize the overall QoS. An algorithm is derived (without the interference constraints of the cognitive radio context) using a Lagrangian relaxation technique to overcome the discrete scheduling constraints. However, a rigorous proof of the convergence and optimality of the proposed algorithm is not provided.

In the cognitive radio context, in which additional interference constraints to protect the primary users must be taken into account, the rate maximization problem was studied in [3, 6, 11]. In these works, the authors consider decentralized solutions in MIMO systems via non-cooperative game theory without spectrum scheduling constraints. The major disadvantage of such decentralized approaches is that the Nash equilibrium solution (i.e., the natural solution concept in non-cooperative games) provides an operating point that is often outperformed by a centralized solution. Other works study rate maximization problems under different CSI assumptions. In particular, [25] addresses the scheduling aspect with partial-CSI at the SU which limits the adaptability to the actual channel state. For a time-varying system, the authors of [26] study dynamic cognitive radio settings without explicit interference temperature constraints imposed by the primary users' presence. Online optimization and no-regret distributed learning algorithms are used in [26] assuming the users do not know the perfect CSI prior to their transmissions. Such a complex approach is not required here, as we consider that CSI is available to the centralized system manager.

Energy efficiency problems in the CR context were studied with QoS and spectrum scheduling constraints in [10], in which the authors minimize the SU's power consumption. The framework in [10] is the closest among all cited references to our paper. However, there is no proof of optimality of the proposed scheduling, which is based on a heuristic method involving some exhaustive search steps. In our work, we use convex optimization tools to find the optimal joint scheduling and power allocation under interferences and QoS constraints. Our optimal solution is calculated via an iterative sub-gradient algorithm that is proven to converge to the optimal solution.

1.2 Our contributions

The main contributions of this paper are summarized here below:

- We derive necessary conditions and sufficient conditions for the existence of a solution to the joint spectrum scheduling and power allocation problem in a CR system.
- We introduce a convex optimization problem based on Lagrangian relaxation of the initial non-convex problem. Then, we prove that the discrete solutions of the relaxed problem are the solutions to the initial problem.
- The optimal solution of the relaxed problem, when it exists, is computed via a projected sub-gradient algorithm. We prove that, when the problem is feasible, our proposed projected sub-gradient algorithm converges to an optimal solution that satisfies the KKT conditions.
- We also study the specific case of a symmetric system for which our iterative algorithm is not suitable and we solve it analytically.
- Numerical results illustrate the energy-efficiency of the proposed allocation strategy compared with other spectrum allocation policies.

The remainder of this paper is organized as follows. The system model is presented in Section 2. In Section 3, we study the joint scheduling and power allocation problem by discussing its feasibility, the Lagrangian relaxation, and its optimality, and then provide a sub-gradient algorithm to solve this problem. Selected numerical results are illustrated in

Section 4 to show the efficiency of the proposed solution. Particular cases, for which an analytical solution is found, are studied in Section 5. Finally, Section 6 concludes the paper.

2. SYSTEM MODEL

We focus on the CR model in Fig. 1 composed of $Q \geq 1$ SUs and $K \geq 1$ PUs. Each primary/secondary user consists of a primary/secondary transmitter (PT/ST) and a primary/secondary receiver (PR/SR), respectively. The transmission is performed over N orthogonal frequency bands. The transmit power of $S\ T_q$ (of the qth SU) in frequency band n∈N≜{1,...,N}n∈N≜{1,...,N} is denoted by p_{qn}, the power allocation of the qth SU is denoted by p−q= (pq1,pq2,...,pqN) ∈RN+p_q=) ∈R+N, ∀q∈Q≜{1,...,Q} ∀q∈Q≜{1,...,Q}, and the overall power allocation profile for all SUs is denoted by $p = (\underline{p}_1, \underline{p}_2, \dots, \underline{p}_Q) \in \mathbb{R}_+^{N \times Q}$.

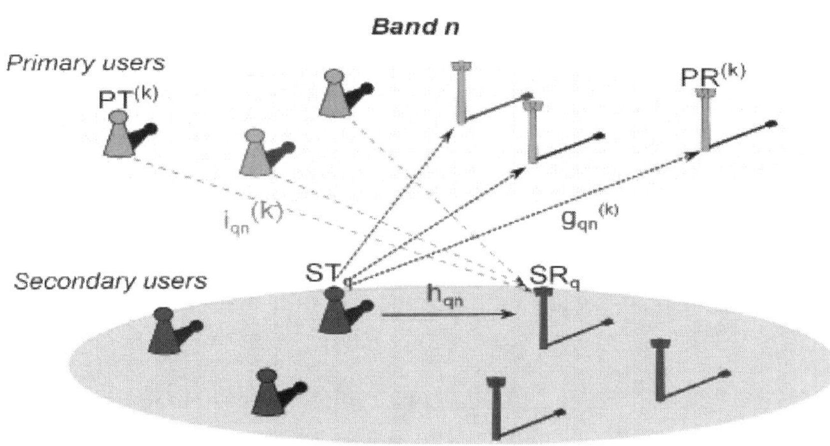

Figure 1. Only $S\ U_q$ is active in band n

The received signal at $S\ R_q$ in band n can be written as:

$$y_{qn} = \sqrt{p_{qn} h_{qn}} v_{qn} + \sum_{k \in \mathcal{K}} i_{qn}^{(k)} + w_{qn}, \tag{1}$$

where h_{qn} is the power gain of the direct link $ST_q - SR_q$; $v_{qn} \backsim \mathcal{CN}(0,1)$ is the normalized transmitted signal by SU q (a zero-mean circularly symmetric complex Gaussian variable of unit variance), $w_{qn} \backsim \mathcal{CN}\left(0, \sigma_{qn}^2\right)$ is the noise in band n for SU_q of variance σ_{qn}^2; and $i_{qn}^{(k)} \backsim \mathcal{CN}\left(0, \left(\tau_{qn}^{(k)}\right)^2\right)$ is the interfering signal from PU k∈ $\mathcal{K} \triangleq \{1,\ldots,K\}$ of variance $\left(\tau_{qn}^{(k)}\right)^2$. Since the transmit power of the PUs cannot be impacted by the secondary system, the terms $i_{qn}^{(k)}$ are just some fixed parameters in our model. Only the knowledge of the SINRs are needed and measured at each SU. All links are assumed to be stationary and independent of the noise.

The Gaussian input, interference and noise assumptions are fairly standard in the signal processing for communications literature [3, 27, 28]. The main scientific reasons behind this are that the Gaussian noise is known to be the worst additive noise distribution in terms of the Shannon achievable rate [29], and that the Gaussian input is optimal in a Gaussian environment [30].

In this context, we define the QoS measure for SU q in band n by the Shannon capacity expression in [31], assuming that the corresponding bandwidth is unitary:

$$c_{qn}(x_{qn}, p_{qn}) = x_{qn} \log_2\left(1 + s_{qn}p_{qn}\right) \tag{2}$$

where s_{qn} is the signal-to-interference-plus-noise-ratio (SINR) of the direct link of qth SU ($ST_q - SR_q$), $s_{qn} = h_{qn} / \left(\sigma_{qn}^2 + \sum_{k=1}^{K}\left(\tau_{qn}^{(k)}\right)^2\right)$. The spectrum allocation policy for the qth SU is denoted by $(x_{q1}, x_{q2}, \ldots, x_{qN})$, $\forall q \in \mathcal{Q}$, and the overall allocation profile for all SUs is denoted by $x = (\underline{x}_1, \underline{x}_2, \ldots, \underline{x}_Q)$.

This work is solely focused on the case in which at most one SU is allocated in each band n to avoid interference impairments (to the PUs and among SUs). This means that only discrete spectrum allocation policies $x_{qn} \in \{0,1\}$ are allowed by the spectrum manager. Assuming orthogonal and unit bandwidth channels, the overall achievable rate of the qth SU transmission is

$$C_q\left(\underline{x}_q,\underline{p}_q\right)=\sum_{n\in\mathcal{N}} c_{qn}(x_{qn},p_{qn}). \tag{3}$$

The assumption that only one SU is allowed in each band is also made for tractability reasons. Indeed, optimizing the system's achievable sum-rate or some other objective under SU power or rate constraints is a very difficult, intractable and non-convex problem in an interference channel model. The achievable rate of one SU C_q is a non-convex function with respect to p–p_, because of the interference terms from the other SUs. To overcome this issue, one possibility is to decentralize the decision process and to consider a distributed cognitive radio system, in which each SU chooses its own power allocation policy to optimize its own objective [3, 6, 11]. The underlying non-cooperative game is a convex game as each SU's rate is convex w.r.t. its own controlled variables $\underline{p}_q$.

However, the major disadvantage of this approach is that the resulting Nash equilibrium solution provides an operating point that performs very poorly compared to a centralized solution. Since we are interested in a centralized cognitive radio system, a second way to overcome this major difficulty (the non - convexity of C_q) is to limit the access of the SUs to the spectrum; only one SU is allowed per band.

Although the presence of PUs impacts the rates of the SUs via the SINR terms s_{qn}, their crucial impact is in the additional interference constraints imposed by their presence on the SUs' transmit powers. Indeed, in our CR system, the SUs are allowed to transmit only if the created interference to the primary receivers is guaranteed to be kept below some predefined thresholds to protect the transmissions of the PUs.

3. JOINT SCHEDULING AND POWER ALLOCATION PROBLEM

The main objective of the paper is to study a centralized resource allocation problem in which the spectrum manager wishes to schedule the SUs in an effort to minimize the overall power consumption (in coherence with the green communications sprint [32, 33]) while meeting minimum QoS constraints at the SUs and without interfering with the PUs above the critical limits. Two main questions arise:

- Spectrum scheduling: which SU should be scheduled in each band?
- Power allocation: what is the optimal power allocation policy for each SU in its allocated bands?

In order to tackle these questions, we formulate the problem as follows:

$$
(DP_1) \quad
\begin{aligned}
&\text{minimize} \sum_{q \in \mathcal{Q}} \sum_{n \in \mathcal{N}} p_{qn} \\
&\text{s.t.} \sum_{n \in \mathcal{N}} g_{qn}^{(k)} p_{qn} \le \overline{P}_q^{(k)}, \ \forall q, \forall k \\
&\quad 0 \le g_{qn}^{(k)} p_{qn} \le P_{qn}^{peak(k)}, \ \forall n, \forall q, \forall k \\
&\quad C_q \left(\underline{x}_q, \underline{p}_q \right) \ge R_q^{min}, \forall q \\
&\quad \sum_q x_{qn} \le 1, \forall n \\
&\quad x_{qn} \in \{0, 1\}, \forall n, \forall q,
\end{aligned}
$$

where R_q^{min} represents the target QoS at SU_q; $g_{qn}^{(k)}$ is the power gain of the interfering link $ST_q - PR^{(k)}$; $\overline{P}_q^{(k)}$ is the maximum average interference power that SU $_q$ is allowed to inflict on PU_k; and $P_{qn}^{peak(k)}$ is the maximum peak interference power in band n that SU $_q$ is allowed to inflict on PU $_k$.

This optimization problem is difficult for two reasons. First, the target QoS constraints and the maximum interference constraints inflicted on the PUs are opposing ones and, thus, the feasible set may be void depending on the system parameters. Second, to avoid the interference impairments to the PUs and among SUs, we assume that the system owner schedules at most one SU to a given band n and that such a band cannot be further fractioned. This turns the problem into a discrete optimization with respect to the scheduling policy. In the remainder of this section, we tackle both issues and provide an efficient algorithm to compute the optimal solution when it exists.

3.1 Feasible set

The spectrum manager has to schedule all SUs to ensure a non-zero QoS target for each SU. These constraints might require the SUs to transmit at power levels which inflict an interference level that is unacceptable by the primary system. Thus, the first arising question is under what conditions on the system parameters are the QoS and the interference constraints met simultaneously?

We denote the feasible set of (DP 1) by SF:

$$\mathcal{S}_{\mathcal{F}} = \left\{ \begin{array}{l} (\boldsymbol{x},\boldsymbol{p}) \in \{0,1\}_{+}^{N\times Q} \times \mathbb{R}_{+}^{N\times Q} : \\ \sum_{n\in\mathcal{N}} g_{qn}^{(k)} p_{qn} \leq \overline{P}_{q}^{(k)}, \forall q, \forall k \\ 0 \leq g_{qn}^{(k)} p_{qn} \leq P_{qn}^{peak(k)}, \forall n, \forall q, \forall k \\ C_q\left(\underline{x}_q, \underline{p}_q\right) \geq R_q^{min}, \forall q \\ \sum_{q\in\mathcal{Q}} x_{qn} \leq 1, \forall n \text{ and } x_{qn} \in \{0,1\}, \forall n, \forall q. \end{array} \right\} \quad (4)$$

We provide next sufficient and necessary conditions on the system parameters for the existence of at least one solution to (DP 1), i.e., SF≠∅.

We start by assuming, without loss of generality, that Rminq>0Rqmin>0, for all SU q. Given this assumption and the fact that only one SU is allowed to transmit in a given band, the problem has no solution when $N<Q$. Thus, a first trivial necessary condition for a solution to exist is that $N\geq Q$. In this case, we derive further necessary conditions and sufficient conditions for the existence of a solution.

Theorem 1.

(Necessary conditions) Assuming that $N \geq Q \text{ and } R_q^{min} > 0, \forall q,$ if the minimum rate Rminℓ$R_\ell$min of an arbitrary SU $\ell$is greater than its maximum achievable rate Rmaxℓ$R_\ell$max, then the feasible set SF is void. Here, Rmaxℓ$R_\ell$max represents the optimal value of the following optimization problem:

$$\begin{aligned} & maximize \sum_{n \in \mathcal{N}} \log_2 (1 + p_{\ell n} s_{\ell n}) \\ & s.t. \ \sum_n g_{\ell n}^{(k)} p_{\ell n} \leq \overline{P}_{\ell}^{(k)} \\ & \quad g_{\ell n}^{(k)} p_{\ell n} \leq P_{\ell n}^{peak(k)}, \ \forall n, \forall k \end{aligned} \tag{5}$$

which corresponds to the maximum achievable rate of SU ℓ if it were the only SU in the network.

This result means that, if there is at least one SU ℓ that cannot achieve its minimum requirement Rminℓ even if it were the only SU in the system, the problem (DP $_1$) is infeasible. In the particular case in which there is only one SU in the system, i.e., Q=1, these necessary conditions are tight as they are also the sufficient conditions guaranteeing the existence of a solution. Although these conditions are not tight when Q>1 (at least one band has to be allocated per SU and, thus, no SU will have access to all bands), they have the merit of being general, fair from the SUs perspective, intuitive, and having a low computational complexity. The proof of this Theorem is detailed in Appendix A1.

Theorem 2.

(Sufficient conditions) Assuming that $N \geq Q$ and $R_q^{min} > 0, \forall q,$ if for each SU q the minimum rate Rmin q is lower than the following threshold: $R_q^{CS} = \log_2 \left(1 + \min_n \left\{ s_{qn} \min_k \left\{ \frac{\overline{P}_q^{(k)}}{g_{qn}}, \frac{P_{qn}^{peak(k)}}{g_{qn}} \right\} \right\} \right),$ then the feasible set is non-void $\mathcal{S}_{\mathcal{F}} \neq \emptyset$.

Intuitively, if each $S\,U_q$ has a minimum rate RminqRqmin small enough, i.e., smaller than the rate obtained when using only its worst channel, then the problem (DP $_1$) is feasible.

Remark 3.1.

When $Q = 1, N = 1,$ the sufficient conditions of Theorem 2 are identical to the necessary conditions of Theorem 1. However, (similarly to the necessary conditions above) these sufficient conditions are not tight in general. Indeed, when Q=1 and N>1 the necessary conditions in Theorem 1 are the tight sufficient conditions. Nevertheless, in the most general case, it seems very difficult to find better sufficient conditions that are computationally tractable and fair with respect to all SUs. To better understand this, consider first the case in which $Q \leq N < 2Q$: there are not enough channels to allocate two channels per SU. Thus, some of the SUs will only be allocated a single channel which may very well be their worst channel. In this case, finding better and fair sufficient conditions seems very unlikely. Now, if $\gamma Q \leq N < (\gamma+1)Q$ with $\gamma \geq 2$: at least $\gamma \geq 2$ channels may be allocated to each SU. In this case, a better sufficient condition could be found by computing the achievable rate over the worst γ channels for each SU. However, since we cannot know in advance which combination of γ channels results in a worst case achievable rate, one would have to compute all CγNCNγ combinations for each of the Q SUs. The complexity of such an approach is therefore prohibitive.

The proof of this Theorem is detailed in Appendix A2.

Now, if the feasible set SFSF is non-void, finding the solution of (DP_1) is not trivial. Indeed, we notice that the scheduling of SUs is a discrete combinatorial problem $x_{qn} \in \{0,1\}$, $\forall n$, $\forall q$. In the following, we will provide an efficient algorithm that computes the optimal solution in a very efficient manner using a Lagrangian relaxation approach [14].

3.2 Lagrangian relaxation

As we have already mentioned, we assume that at most one SU is allowed to transmit in a given band. The constraints on the discrete scheduling policy $\boldsymbol{x}$ make the problem (DP_1) very difficult to solve in this form. In the following, we will use a Lagrangian relaxation technique to overcome this issue.

When the feasible set SFSF is not void, we propose to solve a continuous problem in which the scheduling parameter is continuous $x_{qn} \in [0,1]$. This relaxed problem (CP_2) holds the advantage of being a convex optimization problem, much simpler to solve than (DP_1):

$$(CP_2)\quad \begin{aligned} &\text{minimize} \sum_{q\in\mathcal{Q}}\sum_{n\in\mathcal{N}} p_{qn} \\ &\text{s.t.} \sum_{n\in\mathcal{N}} g_{qn}^{(k)} p_{qn} \le \overline{P}_q^{(k)}, \forall q, \forall k \\ &0 \le g_{qn}^{(k)} p_{qn} \le x_{qn} P_{qn}^{peak(k)}, \forall n, \forall q, \forall k \\ &R_q\left(\underline{x}_q, \underline{p}_q\right) \ge R_q^{min}, \forall q \\ &\sum_q x_{qn} \le 1, \forall n \\ &0 \le x_{qn} \le 1, \forall n, \forall q. \end{aligned}$$

Inspired by [13], the continuous problem (CP $_2$) is not exactly the Lagrangian relaxation of (DP $_1$) but that of an equivalent discrete problem. Two differences can be observed in (CP $_2$). First, the peak interference power constraints $g_{qn}^{(k)} p_{qn} \le P_{qn}^{peak(k)}$ are replaced by $g_{qn}^{(k)} p_{qn} \le x_{qn} P_{qn}^{peak(k)}$. Second, we replace the Shannon achievable rate of SU q by the function:

$$R_q(\underline{x}_q, \underline{p}_q) = \sum_{n\in\mathcal{N}} r_{qn}(x_{qn}, p_{qn}), \tag{6}$$

where

$$r_{qn}(x_{qn}, p_{qn}) = \begin{cases} x_{qn}\log_2\left(1 + \frac{s_{qn} p_{qn}}{x_{qn}}\right), & \text{if } x_{qn} > 0 \\ 0, & \text{otherwise.} \end{cases} \tag{7}$$

Remark 3.2.

Our target problem (DP $_1$) focuses only on discrete spectrum allocation policies $x_{qn} \in \{0,1\}$. Under this assumption, the function in Eq. (7) corresponds exactly to the achievable Shannon rate in (2). However, in the continuous case in which $x_{qn} \in [0,1]$ expression (7) does not correspond to the Shannon achievable rate. The denominator of the term $\frac{s_{qn} p_{qn}}{x_{qn}}$ is introduced for technical purposes and plays an important role in proving the optimality of the Lagrangian relaxation approach when solving (DP $_1$).

The major interest of this approach is that it will allow us to solve the initial non-convex problem using a very efficient algorithm based on convex optimization techniques. Not only is the continuous problem (CP 2) simpler to solve, but also, by using these two modifications, the problem (CP 2) will have discrete solutions in ***x*** which will turn out to be the solutions to our initial problem (DP 1), as we will show later on.

In the following, we will give some proprieties of the problem (*C P* 2) where we denote by (***x*** *,***p*** *) the pair of scheduling and power allocation that satisfies all the constraints at the optimum:

Proposition 1.

In the continuous problem (CP 2), all the rate constraints are active at the optimal solution, i.e., the rate constraints are satisfied with equality.

This result means that if the feasible set is non-void, the optimal rate at each SU achieves the target QoS R^{min}.

Proposition 2.

In the continuous problem (CP 2), all the scheduling average constraints are all active at the optimal solution : ∑qx∗qn=1∑qxqn∗=1, $\forall q$.

This result means that if the feasible set is non-void, all bands are fully used by the opportunistic users. The proofs of these two propositions are detailed in Appendix B.

3.2.1 Dual formulation

The continuous problem (CP 2) is a convex optimization problem. Indeed, the objective function $\sum_q \sum_n p_{qn}$ is affine in the overall power allocation profile p= $(\underline{p}_1, \underline{p}_2, \ldots, \underline{p}_Q) \in \mathbb{R}_+^{N\times Q}$ and, regarding the constraints, the interference constraints are both affine in (*x*,*p*); the scheduling constraints are affine in *x*; and the rate constraint is jointly concave in (*x*,*p*). From the convex optimization problem definition [34], it follows that (CP 2) is convex and, thus, can be solved via a dual formulation. We associate dual variables $\boldsymbol{\lambda} = (\lambda_q)_{q\in\mathcal{Q}}$ with total interference power constraints, $\beta = (\beta_q)_{q\in\mathcal{Q}}$ with rate constraints and $\mu = (\mu)_{n\in\mathcal{N}}$ with scheduling constraints, resulting in the following Lagrangian:

$$L(\boldsymbol{\lambda}, \boldsymbol{\beta}, \boldsymbol{\mu}, \mathbf{x}, \mathbf{p}) = \sum_q \sum_n p_{qn} - \sum_q \beta_q \left(R_q \left(\underline{x}_q, \underline{p}_q \right) - R_q^{min} \right)$$
$$+ \sum_k \sum_q \lambda_q^{(k)} \left(\sum_n g_{qn}^{(k)} p_{qn} - \overline{P}_q^{(k)} \right) + \sum_n \mu_n \left(\sum_q x_{qn} - 1 \right). \quad (8)$$

To solve (CP 2) it suffices to solve:

$$\max_{(\boldsymbol{\lambda},\boldsymbol{\beta},\boldsymbol{\mu}) \geq 0} \min_{(\mathbf{x},\mathbf{p}) \in \mathcal{S}_{\mathcal{F}}} L(\boldsymbol{\lambda}, \boldsymbol{\beta}, \boldsymbol{\mu}, \mathbf{x}, \mathbf{p}). \quad (9)$$

Inspired by [13], in which the authors consider a rate maximization joint scheduling and power allocation problem in OFDM systems without the PU constraints, we propose to solve here a more complex problem (CP_2) via the following steps.

Step 1: The optimal power **p** which minimizes L(**λ**,**β**,**μ**,**x**,**p**) given fixed **λ**, **β** and **μ** [1] is a water-filling type of solution:

$$p_{qn}^* = \frac{x_{qn}}{s_{qn}} \left[\frac{\beta_q s_{qn}}{1 + \sum_k \lambda_q^{(k)} g_{qn}^{(k)}} - 1 \right]_0^{\min_k \left\{ P_{qn}^{peak(k)} \right\} s_{qn}} \quad (10)$$

Step 2: Substituting **p** * into L(**λ**,**β**,**μ**,**x**,**p**) yields to the following affine function in **x**:

$$G(\boldsymbol{\lambda}, \boldsymbol{\beta}, \boldsymbol{\mu}, \mathbf{x}) = - \sum_q \sum_n x_{qn} \left(\varphi_{qn} - \mu_n \right)$$
$$+ \sum_q \beta_q R_q^{min} - \sum_k \sum_q \lambda_q^{(k)} \overline{P}_q^{(k)} - \sum_n \mu_n, \quad (11)$$

where $\varphi_{qn} = \varphi_{qn}(a, b, c, d)$ is a function of the system parameters defined by:

$$\begin{cases} 0 & \text{if } \frac{a}{c} \geq b \\ -b + \frac{a}{c} + b \log_2 \left(\frac{bc}{a} \right) & \text{if } \frac{a}{c} \leq b \leq \frac{a(dc+1)}{c} \\ -da + b \log_2 (1 + dc) & \text{if } \frac{(a)(dc+1)}{c} < b \end{cases} \quad (12)$$

for $a = 1 + \sum_k \lambda_q^{(k)} g_{qn}^{(k)}$, $b = \beta_q$, $c = s_{qn}$ and $d = \min_k \left\{ P_{qn}^{peak(k)} \right\}$.

Step 3: From Eq. (11), we remark that this function is affine in ***x***. Optimizing it over ***x*** such that $x_{qn} \in [0,1]$ yields the dual function:

$$\begin{aligned} G(\lambda, \beta, \mu) = & -\sum_q \sum_n [\varphi_{qn} - \mu_n]^+ + \sum_q \beta_q R_q^{min} \\ & -\sum_k \sum_q \lambda_q^{(k)} \overline{P}_q^{(k)} - \sum_n \mu_n, \end{aligned} \quad (13)$$

where the optimal scheduling allocation ***x*** * is:

$$x_{qn}^*(\mu) = \begin{cases} 1 \text{ if } \varphi_{qn} > \mu_n \\ 0 \text{ if } \varphi_{qn} < \mu_n \end{cases} \quad (14)$$

and if $\varphi_{qn} = \mu_n$, then the optimal value x∗qnxqn∗ can be anything in the interval [0,1]. In such cases and from Proposition 2, one must only ensure that all scheduling constraints are met: $\sum_q x_{qn}^* = 1$ for all *n*.

We can now maximize the dual function G(***λ***,***β***,***μ***) over ***μ*** for given ***λ*** and ***β***, by setting μn=μ∗n(λ,β) similarly to [13] where μ∗n is obtained as follows:

$$\mu_n^*(\lambda, \beta) = \max_q \varphi_{qn}(\lambda_q^{(k)}, \beta_q). \quad (15)$$

Remark 3.3. From Eqs. (*14*) and (*15*), it is clear that . Intuitively, this means that band *n* is allocated to a specific SU *ℓ* if it maximizes a specific channel metric given by . There may be ties when multiple SUs achieve the value μ∗nμn∗ in band *n*. For example, if there exist two SUs *r*, *q* such that. These ties happen with zero probability if the independent random channel gains are drawn from continuous distributions. In practice, this implies that such problematic cases almost never occur.

From Eq. (14) we also observe that (CP $_2$) allows for discrete solutions in ***x*** * when it is feasible. Since we are interested in solving the discrete problem (DP $_1$), we make the choice to select a discrete solution such that

$$x^*_{qn}(\mu^*) = \begin{cases} 1 \text{ if } \varphi_{qn} = \mu^*_n \\ 0 \text{ if } \varphi_{qn} < \mu^*_n. \end{cases} \quad (16)$$

Step 4: Substituting $\boldsymbol{\mu}^*$ into G($\boldsymbol{\lambda},\boldsymbol{\beta},\boldsymbol{\mu}$) and noticing that $\boldsymbol{\mu}^*,\mathbf{p}^*,\mathbf{x}^*$ are all functions of

$$\begin{aligned} G(\boldsymbol{\lambda}, \boldsymbol{\beta}) = &-\sum_n \max_q \varphi_{qn}\left(\lambda_q^{(k)}, \beta_q\right) + \sum_q \beta_q R_q^{min} \\ &-\sum_k \sum_q \lambda_q^{(k)} \overline{P}_q^{(k)}. \end{aligned} \quad (17)$$

The solution to the dual problem (9) can be computed numerically by maximizing G($\boldsymbol{\lambda},\boldsymbol{\beta}$) over $\boldsymbol{\lambda} \geq 0$ and $\boldsymbol{\beta} \geq 0$. To this aim, we use a sub-gradient based search [14] to update $\boldsymbol{\lambda}$ and $\boldsymbol{\beta}$.

3.2.2 Projected sub-gradient algorithm

Based on the previous dual formulation, we propose an iterative sub-gradient algorithm to compute the solution of (CP_2) when it exists. The iterations are detailed in Algorithm 1. The sub-gradient approach is usually employed to compute water-filling type of solutions [3, 10, 13]. The proposed algorithm converges to the optimal solution whenever it is feasible.

Other efficient algorithms such as interior point methods, which are based on Newton's iteration and on second-order derivatives, can be implemented to solve (CP_2). Both types of algorithms have advantages and inconveniences, and their convergence performance depends on the compromise between the number of iterations and the complexity of each iteration. The sub-gradient method has the advantage of being simple as its iterations are of low-complexity and scale as $O(Q \times (K+N))$.

Proposition 3.

If (CP_2) is feasible, then Algorithm 1 always converges and the convergence point is an optimal scheduling and power allocation policy.

The proof of this proposition is based on convex optimization tools and is given in Appendix C. The intuition is that, at the convergence state, the iteration, which is discrete with respect to the spectrum allocation policy by construction, satisfies the KKT conditions which are both necessary and sufficient for optimality in convex optimization problems.

Algorithm 1: Projected Sub-gradient Algorithm for the Joint Scheduling and Power Allocation Problem

---

1) Initialization: $\beta_q^{[0]}, \lambda_q^{(k)[0]}, \forall q, \forall k$
2) $t = 1$
3) while *iteration t* **do**
  for *each SU* $q \in \mathcal{Q}$ **do**
    for *each Band* $n \in \mathcal{N}$ **do**
      Update powers $p_{qn}^{[t]}$ given in (10)
      Update channel metrics φ_{qn} defined in (12)
      Update the Lagrange multiplier $\mu_n^{[t]}$ using (15)
      Update the scheduling allocation $x_{qn}^{[t]}$ using (16)
    end
    for $k \in \mathcal{K}$ **do**
      Update $\lambda_q^{(k)[t]}$ such that[2]:

$$\lambda_q^{(k)[t]} = \left[\lambda_q^{(k)[t-1]} - \tau\left(\overline{P}_q^{(k)} - \sum_n g_{qn}^{(k)} p_{qn}^{[t]}\right)\right]^+$$

    end
    Update $\beta_q^{[t]}$ such that:

$$\beta_q^{[t]} = \left[\beta_q^{[t-1]} + \tau\left(R_q^{min} - R_q\left(\underline{x}_q^{[t]}, \underline{p}_q^{[t]}\right)\right)\right]^+$$

    where R_q is defined in (6)
    $t = t + 1$
  end
end
4) Repeat **3)** until convergence for all SUs in all the bands $\forall q, \forall n$, $\left|x_{qn}^{[t]} - x_{qn}^{[t-1]}\right| \leq \epsilon_x$, $\left|p_{qn}^{[t]} - p_{qn}^{[t-1]}\right| \leq \epsilon_p$, $\left|\beta_q^{[t]} - \beta_q^{[t-1]}\right| \leq \epsilon_\beta$ and
$\max_{k \in \mathcal{K}} \left\{\left|\lambda_q^{(k)[t]} - \lambda_q^{(k)[t-1]}\right| - \epsilon_{\lambda_k}\right\} \leq 0$
where ϵ_x, ϵ_p, ϵ_{λ_k} and ϵ_β are precision parameters.

3.3 Optimality of the Lagrange relaxation

Our main objective in the remainder of the paper is to show that Algorithm 1 not only converges to the solution of (CP $_2$) but to that of the initial discrete problem (DP $_1$).

Let us define by DP $_2$ the discrete version of (CP $_2$). The objective of DP $_2$ is to minimize the SU overall power consumption subject to the rate constraints, the interference power constraints (identical to the ones in (CP $_2$)) and the discrete scheduling constraints: ∀n. Because of space limitations, we do not write the expressions of DP $_2$ explicitly.

Proposition 4.

If (CP $_2$) is feasible, then all the discrete solutions of (CP $_2$) (with respect to the optimal scheduling allocation $\boldsymbol{x}^*$) are the only optimal solutions of DP $_2$.

Proof.

The feasible set of DP $_2$ is included in the feasible set of (CP $_2$). The only difference between both problems is the scheduling parameter x_{qn} which is discrete in DP $_2$ and continuous in (CP $_2$). On the one hand, we can see that if there are any discrete solutions to (CP $_2$), then these discrete solutions will also be the optimal solutions of DP $_2$. On the other hand, if the solutions of (CP $_2$) are continuous (i.e., the scheduling allocations are between 0 and 1), then the SUs' power allocations policies are constants (as they do not depend on the scheduling allocations), thus, these continuous solutions have the same efficiency comparing to the discrete solutions on the borders (i.e., are equal to 0 or 1).

Next, we compare the discrete problems (DP $_1$) and DP $_2$.

Proposition 5.

The optimization problems (DP $_1$) and DP $_2$ are equivalent in the sense that their solution sets are identical.

Proof.

If we fix an arbitrary scheduling policy x˜∈Xx~∈X, where the scheduling constraints set is defined as then, both problems (DP $_1$) and DP $_2$ reduce to the same power allocation problem (CP 3) below in which the only variable left is the power allocation policy[3] at the SU's

$$(CP_3)\quad \begin{aligned} &\text{minimize} \sum_{q\in Q}\sum_{n\in N} p_{qn} \mathbb{1}_{\{\tilde{x}_{qn}\}} \\ &\text{s.t.} \sum_{n} g_{qn} p_{qn} \mathbb{1}_{\{\tilde{x}_{qn}\}} \le \overline{P}_q, \ \forall q \in Q \\ &0 \le g_{qn} p_{qn} \mathbb{1}_{\{\tilde{x}_{qn}\}} \le P_{qn}^{peak} \mathbb{1}_{\{\tilde{x}_{qn}\}}, \forall q, n \\ &\sum_{q} \mathbb{1}_{\{\tilde{x}_{qn}\}} \log_2\left(1 + \frac{s_{qn} p_{qn}}{\tilde{x}_{qn}}\right) \ge R_q^{min}, \ \forall n. \end{aligned}$$

On the one hand, if the scheduling parameter equals, then the corresponding optimal power is also zero qn=0 (because the optimal power is linear in **x**). On the other hand, if the scheduling parameter equals one,, then the optimal powers p˜qn are given by a water-filling type of solution over the allocated spectrum. Thus, whenever x is fixed, both problems (DP 1) and DP 2 reduce to solving the same *Q* decoupled power allocation problems.

It remains to prove that both DP 2 and (DP 1) have the same set of optimal spectrum allocation policies ***x*** *. The optimal powers ***p*** * will follow by solving (CP 3). Assuming that (DP 1) is feasible, we can easily check that the optimal solutions of (DP 1) meet all the constraints of DP 2 as well. Adding the fact that the feasible set of (DP 1) contains (and is larger) than the feasible set of DP 2, we conclude that the optimal solution sets of the two problems are identical.

Theorem 3.

If (CP 2) is feasible, then all solution of (CP 2) that are discrete w.r.t ***x*** * are the optimal solutions of (DP 1).

This Theorem follows from Propositions 4 and 5. In Proposition 5, we have shown that both discrete problems DP 2 and (DP 1) are equivalent. Moreover, from Proposition 4, we have shown that the optimal discrete solutions of DP 2 are the discrete solutions of (CP 2). Thus, we can conclude that all discrete solutions of (CP 2) are the optimal discrete solutions of (DP 1).

Corollary 1.

The feasibility of the relaxed problem (*C P* 2) implies the feasibility of (*D P* 1).

This is important because the feasibility of ($C\ P_2$) problem, which is a convex problem, is much simpler to study than the feasibility of the discrete problem ($D\ P_1$). In other words, to decide whether the initial problem ($D\ P_1$) is convex, we have simply to solve the problem of feasibility of ($C\ P_2$) which is convex.

In conclusion, our results show that Algorithm 1, initially built to solve (CP_2), selects only the optimal discrete solutions and actually solves DP_2. From Proposition 3, both discrete problems DP_2 and (DP_1) are equivalent; thus, our projected sub-gradient Algorithm 1 solves the initial problem (DP_1) and converges to the optimal scheduling and power allocation policy whenever the problem is feasible.

In the following section, we will present some simulation results which illustrate the performance of our projected sub-gradient algorithm.

4. NUMERICAL RESULTS

All observations in this section have been verified via extensive simulations with generic system parameters. We have selected only a few of the most illustrative and interesting results to be presented next.

4.1 Power consumption efficiency

We start by comparing the overall power consumption between our optimal scheduling policy and arbitrarily scheduling techniques such as interleaved and block-wise spectrum scheduling [35]. Once the spectrum scheduling is fixed, computing the power allocation policies follows via water-filling type of sub-gradient methods. We focus only on the cases in which the problem is feasible and in which our Algorithm 1 converges to the optimal solution of (DP_1). Thus, for each of 10^4 random channel realizations, we choose the minimum target QoS that are equal to the rates in the sufficient conditions of Theorem 2.

Figure 2 illustrates this comparison as function of $N \in \{Q,2Q,4Q,8Q,16Q\}$ in the scenario: [4]

We remark that the interleaved and block-wise allocations have the same average performance because of the independent and identically distributed (i.i.d) channel gains. Our algorithm outperforms these two fixed-spectrum allocations (interleaved or block-wise allocation), and the performance gap decreases with N.

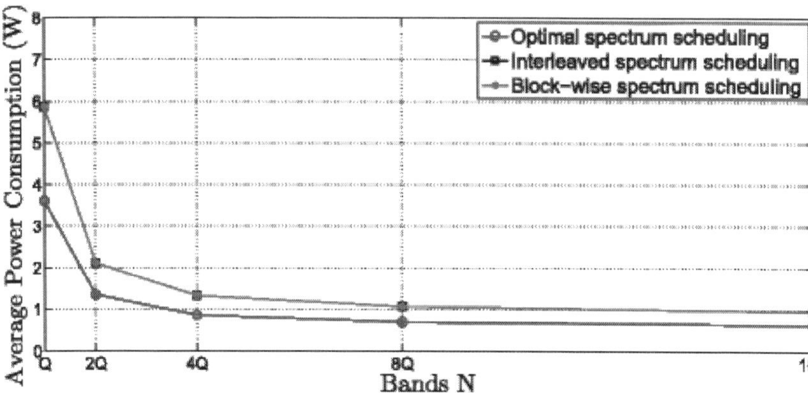

Figure 2. Average power consumption vs. number of bands $N \in \{Q, 2Q, 4Q, 8Q, 16Q\}$ for Q=4 when the sufficient conditions are always met. Our optimal spectrum scheduling outperforms the two others spectrum scheduling allocations (interleaved and block-wise)

The sufficient conditions guaranteeing the existence of a solution and the convergence of Algorithm 1 are not tight in general as they rely on the use of Q bands alone and on the assumption that each SU is allowed to transmit in its worst channel. Finding tighter sufficient conditions that are tractable seems a very difficult task. Therefore, we study the performance of our algorithm in the cases in which the sufficient conditions are not met.

4.2 Problem feasibility

Figure 3 illustrates the empirical probability (over 10^4 random channels realizations) that the problem (DP $_1$) is not feasible as function of the target QoS R^{min} in the following scenario: $N = 4, K = 8, Q = 2, R_q^{min} = R^{min}, \forall q, \overline{P}_q^{(k)} =$ 10 mW$, \forall q, \forall k, s = \begin{pmatrix} 27.8797 & 2.0727 & 0.7779 & 4.3263 \\ 7.6688 & 7.6722 & 11.0049 & 1.7281 \end{pmatrix}^4,$ $P_{qn}^{peak(k)} = 20$ mW$, \forall q, \forall n, \forall k$ and the interfering channel gains are drawn randomly such that $g_{qn}^{(k)} \backsim \mathcal{N}(0,\sigma_g^2), \forall q, \forall n, \forall k$ for an arbitrary chosen value $\sigma_g^2 = 2$.

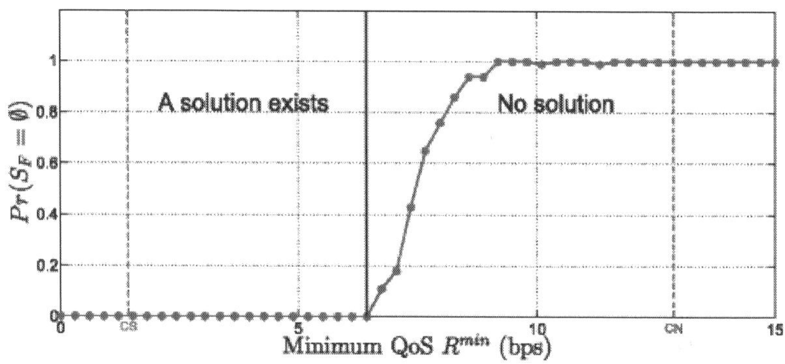

Figure3. Empirical probability that the problem is not feasible as function of the minimum target rate R^{min} for each SU

In order to decide whether the problem is feasible or not, we test if the algorithm converges before the maximum number of iterations is reached. From Proposition 3 and Theorem 3, if the algorithm converges, then the convergence point is an optimal solution of (DP $_1$). For practical reasons, if the algorithm reaches the maximum number of iterations before convergence, we decide that the problem is not feasible. This approach is based on an empirical search for the maximum number of iterations. In this setting, we fix 10^5 maximum number of iterations.

In Fig. 3, we remark that there is a threshold, $R^{min} \simeq 6.25$ bps, below which a solution exists and above which the problem is not feasible. The other two plotted thresholds, CS and CN, illustrate the worst case sufficient conditions and necessary conditions over all random realizations and are computed as follows. We denote by the sufficient condition rates of Theorem 2 for the random channel realization $t \in \{1,...,10^4\}$. Similarly, we denote the necessary condition rates by of Theorem 1 which also depend on the random realization t. The threshold CS represents the minimum value of the rates RCS (t)q over all SUs and over all random channel realizations: bps and the threshold CN represents the maximum value of the rates Rmax(t)qRqmax(t) over SUs and over the random channel realizations: maxq, bps.

Although the CS and CN values in Fig. 3 are worst case conditions (over all random channels and all SUs), they still show that our sufficient and necessary conditions are not tight in general, as discussed in Section 3.1. This means that there are a lot of cases in which we do not know a priori whether the problem is feasible or not: all cases in which sufficient conditions are not met but the necessary conditions are met.

4.3 Necessary conditions are not met

We illustrate now the case in which the necessary conditions are not met, i.e., when there exists at least one SUq such that the target rate is above the maximum achievable rate in bps which are higher than the maximum rates that each SU could achieve if they were alone in the system: Rmax1=8.0803 bps, Rmax2=12.8686 bps.

Figure 4 illustrates the evolution of the Lagrange multipliers μ_n and β_q corresponding to the average constraints over the algorithm's iterations. We remark, that Algorithm 1 does not converge before the maximum number of iterations is reached. In this case, we decide that the problem is not feasible. An alternative would be to further increase the maximum number of iterations, but this process has to be finite and a pragmatic decision has to be made at some point.

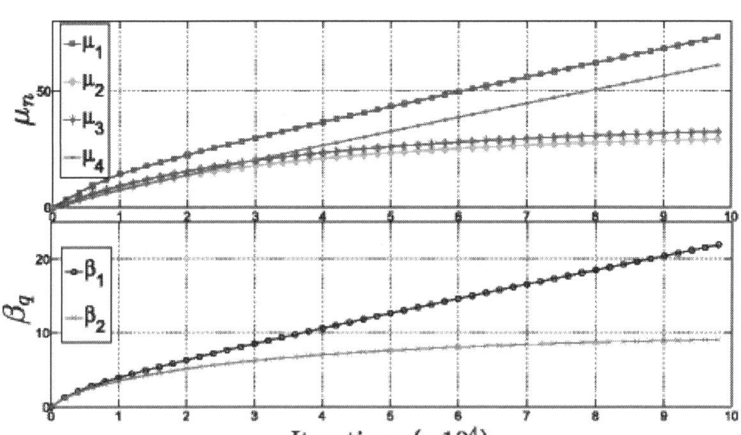

Figure 4. Algorithm 1 reaches the maximum number of iterations (10^5) before convergence

4.4 Sufficient conditions are not met although necessary conditions are met

Here, we illustrate the case in which the necessary conditions are met and at least one of the sufficient conditions is not met in the same setting as Fig. 4 except for the target rates which are in between the thresholds of the sufficient conditions and the necessary conditions.

4.4.1 Convergence of Algorithm 1

In Fig. 5, we want to check whether our projected sub-gradient algorithm converges to an optimal solution or not.

In Fig. 5 a, we plot the Lagrange multipliers μ_n, β_q of the scheduling and target rate constraints when the minimum target rates are bps. We can see that our algorithm converges in this case and, thus, the problem is feasible and solved via Algorithm 1. Indeed, the Lagrange multipliers converge to the optimal values μ∗nμn∗ and β∗qβq∗ that are strictly positive, and the scheduling and rate constraint are satisfied with equality.

In Fig. 5 b, we fix the minimum target rates. In this case, we remark that our algorithm reaches the maximum number of iterations. Because of such high target rates, our algorithm does not converge and, thus, the problem has no solution.

4.5 Water-filling solution

Figure 6 illustrates the optimal scheduling and power allocation policies in the case in which our algorithm converges for the same scenario as in Fig. 5 a.

We remark that the optimal solution is fair in terms of bands per SU; two bands are allocated to each SU. Band 1 and 4 are allocated to SU_1, since SU_1 has higher SNR than SU_2 in these bands. Band 2 and 3 are allocated to SU_2 for the same reason. This power and spectrum allocation policy satisfies all KKT conditions, thus, it is the optimal solution.

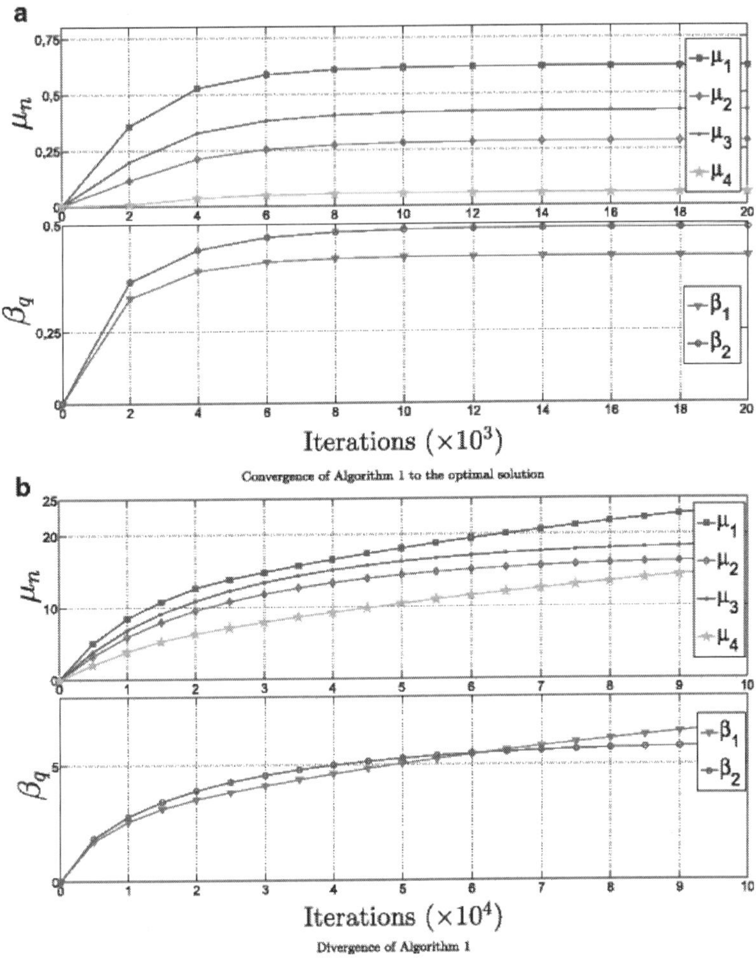

Figure 5. Behaviour of Algorithm 1 when sufficient conditions are not met in Scenario 2: (**a**) R^{min} = [3 3] the problem is feasible and the algorithm converges and (**b**) R^{min} = [8 8] the algorithm does not converge in this case

In conclusion, when the sufficient conditions are not met although the necessary conditions are met, we do not know a priori if Algorithm 1 converges or not. Nevertheless, we can still exploit our algorithm to decide whether the problem is feasible or not and to compute the optimal solution when it exists. If the algorithm converges before reaching the maximum number of iterations, then we know that the problem is feasible and that the convergence point is an optimal solution to (DP 1). Otherwise, we can increase the maximum number of iterations or decide that the problem is not feasible and has no solution. In such cases, instead of not allowing any SU to transmit, the spectrum manager (instead of not allowing any opportunistic user to transmit) can decrease the minimum SU target rates (lower QoS may be better than no QoS in low data rate applications) or may even decide to schedule only a subset of the SUs and turn the problem into a feasible one [8].

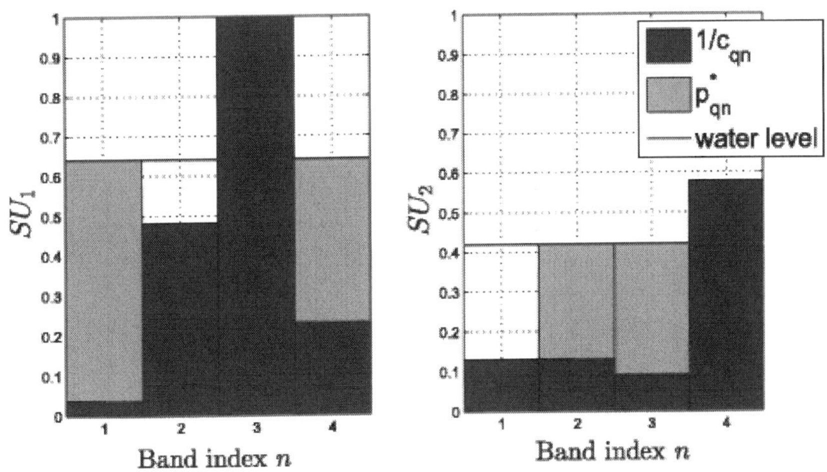

Figure 6. Optimal scheduling and power allocation when Algorithm 1 converges. The system owner allocates two bands per SU

As we have seen before, there exist other cases in which the Algorithm 1 does not calculate the solution. This happens in the indecision cases between several SUs (i.e, to decide whether the SU is allowed to transmit in a given band or not). In the following, we will try to solve some of these specific cases.

5. PARTICULAR CASE: SYMMETRIC SYSTEMS

In this section, we are interested in a specific symmetric system in which Algorithm 1 is not suitable to compute the solution. This happens whenever ties arise between several SUs when deciding on the SU to be allocated a given band: if several SUs have the same maximum value for the decision parameter μ∗n described in Section3.2.1 (there exist two SUs *q*≠*r* such that $R_q^{min} = R^{min}, \forall n, \forall q.$

Consider the completely symmetric system in which all SUs experience the same channel gains, the same peak and total interference constraints, and have the same target rates: $s_{qn} = s, g_{qn} = g, P_{qn}^{peak} = P^{peak}, \overline{P}_q = \overline{P}$ and $R_q^{min} = R^{min}, \forall n, \forall q.$

Remark 5.1.

In this case, the optimal power allocation is such that every SU uniformly allocates its power over its allocated bands. For any scheduling policy ***x***, the optimal power allocation is a water-filling type solution [11, 12] derived from the KKT conditions:

$$p_{qn}^* = \begin{cases} \frac{1}{s}\left[\frac{\beta_q s}{1+\lambda_q g} - 1\right]_0^{P^{peak} s} & if\, x_{qn}^* = 1 \\ 0 & otherwise \end{cases} \tag{18}$$

and, thus, we have for any band *n* that is used at the optimal solution.

In the following, the main idea is to simplify the problem (DP $_1$) in which the unknowns are ***x*** and ***p*** and reduce it into a problem in which the only unknown is denotes the number of bands allocated at the optimum to the SU $_q$ [6] :

From Proposition 2, since the scheduling constraint is satisfied with equality, the overall number of bands allocated to the SUs, at the solution, equals the total number of bands *N*.

$$\sum_q N_q^* = N. \tag{19}$$

According to Proposition 1 (rate constraint is met with equality) and Remark 5.1 (uniform power allocation is optimal), we can write the optimal power as:

$$p_{qn}^{*} = \begin{cases} \frac{1}{s}\left(\exp\left(\frac{R^{min}}{N_q}\right) - 1\right) & \text{if } x_{qn}^{*} = 1 \\ 0 & \text{otherwise.} \end{cases}$$

Using (19), the discrete problem (DP $_1$) simplifies as follows:

$$(DP_N) \quad \begin{aligned} &\text{minimize} \sum_{q \in Q} \frac{N_q}{s}\left(\exp\left(\frac{R^{min}}{N_q}\right) - 1\right) \\ &\text{s.t.} \ \frac{N_q}{s}\left(\exp\left(\frac{R^{min}}{N_q}\right) - 1\right) \leq \frac{\overline{P}}{g}, \forall q \quad (C1) \\ &\frac{1}{s}\left(\exp\left(\frac{R^{min}}{N_q}\right) - 1\right) \leq \frac{P^{peak}}{g}, \forall q \quad (C2) \\ &\sum_{q=1}^{Q} N_q = N \quad (C3) \\ &N_q \in \mathbb{N}^{*}, \forall q. \quad (C4) \end{aligned}$$

This problem is interesting as it is no longer a joint scheduling and power allocation problem $(\boldsymbol{x},\boldsymbol{p})$ and depends only on the number of allocated bands at the optimum N∗q, ∀qNq∗, ∀q.

For simplicity reasons, we start by assuming that the number of available bands is proportional to the number of SUs. This means that the ratio N/Q is an integer: $N/Q \in \mathbb{N}$. .

Now, we will prove that the optimal solution to problem (DP $_N$) is to allocate the spectrum in a fair way to the SUs.

Proposition 6.

The optimal solution of the problem (DP $_N$), when feasible, is to uniformly allocate the spectrum to the SUs:

$$N_q^{*} = \frac{N}{Q}, \forall q. \tag{20}$$

The proof of this Proposition is detailed in Appendix D. The closed-form solution of the power allocation policy is given by:

$$p_{qn}^{*} = \begin{cases} \frac{1}{s}\left(\exp\left(\frac{R^{min}Q}{N}\right) - 1\right) & \text{if } x_{qn}^{*} = 1 \\ 0 & \text{otherwise} \end{cases} \tag{21}$$

where {x*qn}qn is any spectrum allocation policy such that every band is used and each SU is allocated exactly NQNQ bands.

In conclusion, in order to minimize the power consumption of the CR symmetric system, all the spectrum has to be used (see Proposition 1). Since all SUs have the same channel conditions, the spectrum manager does not privilege any particular SU and the spectrum is divided equally among them. There are many ways to allocate the spectrum either by an interleaved or block-wise allocation [35]. These two types of allocations are all equivalent given the symmetry of the system. Then, each SU allocates uniformly its power as in (21) over its allocated bands.

In the following Table 1, we compare the optimal value of the objective function in different particular cases regarding the scheduling policy to illustrate that the fair spectrum allocation among the different users is optimal. We consider the following scenario, which falls into the hypothesis of Proposition is chosen randomly such that we choose *T*=3 and the SINR *s*=2. Notice that all permutations of are equivalent by symmetry and lead to an identical power consumption; therefore, we provide only one of these permutations in the Table. We remind that the sum of the allocated bands must be equal to N i.e. $\sum_q N_q = N$.

Table 1. Comparing different spectrum allocations

SU1	SU2	SU3	$\sum_{q,n} p^*_{qn}$ (mW)
$\frac{N}{Q}$	$\frac{N}{Q}$	$\frac{N}{Q}$	2.5774
$\frac{N}{Q}+1$	$\frac{N}{Q}-1$	$\frac{N}{Q}$	3.1585
$\frac{N-1}{2}$	$\frac{N-1}{2}$	1	10.6598
$N-1-T$	T	1	10.8130
$N-2$	1	1	19.3531

We remark that exactly NQNQ bands are allocated to each SU. This numerical result validates our Proposition 6.

Next, we provide a more general solution, in the case in which the number of bands is not proportional to the number of SUs, $N/Q \notin \mathbb{N}$. . It turns out that the optimal solution to problem (DP $_N$) is very similar to the fair spectrum allocation discussed above.

Proposition 7.

The optimal solution of the problem (DP $_N$) when feasible is given as follows:

$$N_q^* = \begin{cases} \lfloor \frac{N}{Q} \rfloor + 1 & \text{if } q \in \left\{1, \ldots, N - \lfloor \frac{N}{Q} \rfloor Q\right\} \\ \lfloor \frac{N}{Q} \rfloor & \text{if } q \in \{N - \lfloor \frac{N}{Q} \rfloor Q + 1, \ldots, Q\} \end{cases} \tag{22}$$

where $\lfloor y \rfloor$ denotes the integer part of y. The proof of this Proposition is detailed in Appendix D. The optimal power allocation is uniform over all bands that are used by the SUs and is given by (21). The difference with the previous case lies in the number of bands that each SU can use.

In conclusion, when solving the (DP $_N$) problem in this symmetric system and because of the convex shape of the objective function in a continuous space, two cases arise: (1) either the solution to the relaxed convex continuous problem (i.e., N/Q) is integer and the fair spectrum allocation is optimal (2) or the solution NQNQ is not an integer and each SU is allocated either bands such that . This means that the solution is not perfectly fair (as the previous case) and a subset of SUs will be allocated an additional band.

Aside from this particular case (w.r.t. the system parameters), there are other interesting cases but more complicated to solve: (1) the case in which a *SUq* has the same channel gains and peak interference thresholds over all the frequency bands, but two different SUs do not necessary have the same channel conditions: $s_{qn} = s_q$, $g_{qn} = g_q$, Ppeakqn=PpeakqPqnpeak=Pqpeak, $\forall n$; (2) the case in which in a band n all SUs have the same channel gains and peak interference thresholds: $s_{qn} = s_n$, $g_{qn} = g_n$, Ppeakqn=PpeaknPqnpeak=Pnpeak, $\forall q$. Although, in the first case (1), we can simplify the problem using the optimal power uniform allocation for each *SUq*, an analytical solution seems difficult or even impossible to find since, as opposed to the perfectly symmetric case, we can no longer conclude from the KKT conditions that the number of allocated bands is the same for all SUs. In the second case, we can no longer use the fact that the optimal power allocation is uniform over the allocated spectrum, so we cannot even simplify the problem as the symmetric case. Since these cases seem difficult and arise with probability zero in practice, we will not detail them here.

6. CONCLUSIONS

In this work, we have investigated the usage of full CSI by a cognitive radio manager to jointly schedule spectrum access and power allocation for opportunistic users in a power-efficient CR network. Two main challenges emerge in the underlying optimization problem. A first difficulty lies in the QoS and interference power constraints which may not be met simultaneously. To tackle this issue, we have provided general necessary conditions and sufficient conditions for the existence of an optimal solution.

A second challenge lies in the non-convexity of this problem because of the discrete scheduling policies. This aspect is overcome by exploiting a specific Lagrangian relaxation technique. We have proposed an iterative projected sub-gradient algorithm converging to the optimal joint power and spectrum policy whenever the problem is feasible and the channels are asymmetric. It turns out, that the convergence point is the optimal solution to the initial discrete problem. We have also studied the particular case of a symmetric CR network for which a closed-form solution is found.

Future work may consider further analysis of the cases in which the problem is unfeasible. Instead of not scheduling any SU, the radio-resource manager may decide to remove a subset of SUs chosen (similarly to [8]) to make the optimization problem feasible with a limited outage probability.

REFERENCES

1. Spectrum Policy Task Force, Report of the spectrum rights and responsibilities working group, tech. rep.,Federal Communications Commission (2002).

2. JM Peha, Sharing spectrum through spectrum policy reform and cognitive radio. Proc. IEEE. **97**(4), 708–719 (2009).

3. G Scutari, DP Palomar, MIMO cognitive radio: a game theoretical approach. IEEE Trans. Signal Process. **58**(2), 761–780 (2010).

4. D Lee, et al, Coordinated multipoint transmission and reception in LTE-advanced: deployment scenarios and opertaional challenges. IEEE Commun. Mag. **50**(2), 148–155 (2012).

5. KI Pedersen, F Frederiksen, C Rosa, H Nguyen, LGU Garcia, Y Wang, Carrier aggregation for LTE-advanced: functionality and performance aspects. IEEE Commun. Mag. **49**(6), 89–95 (2011).

6. G J-Sh Pang, F Scutari, Ch Facchinei, Wang, Distributed power allocation with rate constraints in gaussian parallel interference channels. IEEE Trans. Inf. Theory. **54**(8), 3471–3489 (2008).

7. R Cavalcante, et al, Toward energy-efficient 5G wireless communications technologies: tools for decoupling the scaling of networks from the growth of operating power. IEEE Signal Proc. Mag. **31**(6), 24–34 (2014).

8. A Alsawah, I Fijalkow, Practical radio link resource allocation for fair QoS-provision on OFDMA downlink with partial channel-state information. EURASIP J. Adv. Signal Process. **2009**(1), 1–16 (2008).

9. D Gesbert, MS Alouini. Selective multi-user diversity, Internaltional Symposium on Signal Processing and Information Technology (Germany, 2003), pp. 162–165.

10. DT Ngo, T Le-Ngoc, Distributed resource allocation for cognitive radio networks with spectrum-sharing constraints. IEEE Trans. Veh. Technol. **60**(7), 3436–3449 (2011).

11. G Scutari, DP Palomar, S Barbarossa, Asynchronous iterative water-filling for Gaussian frequency-selective interference channels. IEEE Trans. Inf. Theory. **54**(7), 2868–2878 (2008).

12. D Palomar, J Fonollosa, Practical algorithms for a family of waterfilling solutions. IEEE Trans. Signal Process. **53**(2), 686–695 (2005).

13. J Huang, VG Subramanian, R Agrawal, RA Berry, Joint scheduling and resource allocation in uplink OFDM systems for broadband wireless access networks. IEEE JSAC. **27**(2), 226–234 (2009).

14. S Boyd, L Xiao, A Mutapcic, *Subgradient Methods*, vol. 2004 (Notes for EE392o, Stanford University, Autumn Quarter, 2003).

15. DD Yu, Cioffi JM, Spc10-2: iterative water-filling for optimal resource allocation in ofdm multiple-access and broadcast channels. IEEE Glob. Telecommun. Conf (2006).

16. G Scutari, DP Palomar, S Barbarossa, Optimal linear precoding strategies for wideband noncooperative systems based on game theory–part I: Nash equilibria. IEEE Trans. Signal Process. **56**(3), 1230–1249 (2008).

17. G Scutari, DP Palomar, S Barbarossa, Optimal linear precoding strategies for wideband non-cooperative systems based on game theory–part II: algorithms. IEEE Trans. Signal Process. **56**(3), 1250–1267 (2008).

18. EV Belmega, S Lasaulce, M Debbah, Power allocation games for MIMO multiple access channels with coordination. IEEE Trans. Wirel. Commun. **8**(6), 3182–3192 (2009).

19. G Miao, N Himayat, YG Li, Crosslayer optimization for energy efficient wireless communications: a survey. Wirel. Commun. Mob. Comput. **9**(4), 529–542 (2009).

20. Ch Isheden, Z Chong, E Jorswieck, Framework for link-level energy efficiency optimization with informed transmitter. IEEE Trans. Wirel. Commun. **11**(8), 2946–2957 (2012).

21. G Miao, N Himayat, YG Li, Distributed interference-aware energy-efficient power optimization. IEEE IEEE Trans. Wirel. Commun. **10**(4), 1323–1333 (2011).

22. G Bacci, EV Belmega, L Sanguinetti, Distributed energy-efficient power optimization in cellular relay networks with minimum rate constraints. IEEE Int. Conf. Acoust. Speech Signal Process. (ICASSP), 7014–7018 (2014).

23. VG Subramanian, RA Berry, R Agrawal, Joint scheduling and resource allocation in CDMA systems. IEEE Trans. Inf. Theory. **56**(5), 2416–2432 (2010).

24. J Huang, VG Subramanian, R Agrawal, RA Berry, Downlink scheduling and resource allocation for OFDM systems. IEEE Trans. Wirel. Commun. **8**(1), 288–296 (2009).

25. A Bagayoko, I Fijalkow, P Tortelier, Power control of spectrum-sharing in fading environment with partial channel state information. IEEE Trans. Signal Process. **59**(5), 2244–2256 (2011).

26. P Mertikopoulos, EV Belmega, Transmit without regrets: online optimization in MIMO-OFDM cognitive radio systems. IEEE J. Sel. Areas Commun. **32**(11), 1987–1999 (2014).

27. R Masmoudi, EV Belmega, I Fijalkow, N Sellami, in *International Symposium DySPAN*. A closed-form solution to the power minimization problem over two orthogonal frequency bands under QoS and cognitive radio interference constraints, pp. 212–222.

28. R Masmoudi, EV Belmega, I Fijalkow, N Sellami, A unifying view on energy-efficiency metrics in cognitive radio channels. Eur. Signal Process. Conf, 171–175 (2014).

29. SN Diggavi, TM Cover, The worst additive noise under a covariance constraint. IEEE Trans. Inf. Theory. **47**(7), 3072–3081 (2001).

30. E Telatar, Capacity of multi-antenna Gaussian channels.European Trans. Telecommun. Wiley Online Libr. **10**(6), 585–595 (1999).

31. CE Shannon, Communication in the presence of noise. Proc IRE. **37**(1), 10–21 (1949).

32. E Masera, T Lestable, *Econhome project presentation* (Mobile World Congress, Barcelona, 2011).

33. H Congzheng, T Harrold, et al, Green radio: radio techniques to enable energy-efficient wireless networks. IEEE Commun. Mag. **49**(6), 46–54 (2011).

34. S Boyd, L Vandenberghe, *Convex Optimization* (Cambridge Univ. Press, Cambridge, 2004).

35. AM Masucci, EV Belmega, I Fijalkow, *Optimal blockwise subcarrier allocation policies in single carrier FDMA uplink systems*, (2014).

36. S Foucart, H Rauhut, *A mathematical introduction to compressive sensing* (Birkhäuser, Boston, 2013).

37. Nesterov Y, Introductory Lectures on Convex Programming Volume I: Basic course, Lecture notes (1998).

38. NZ Shor, *Minimization methods for non-differentiable functions*, vol. 3 (Springer-Science & Business Media, 2012).

INDEX